W9-BZJ-437

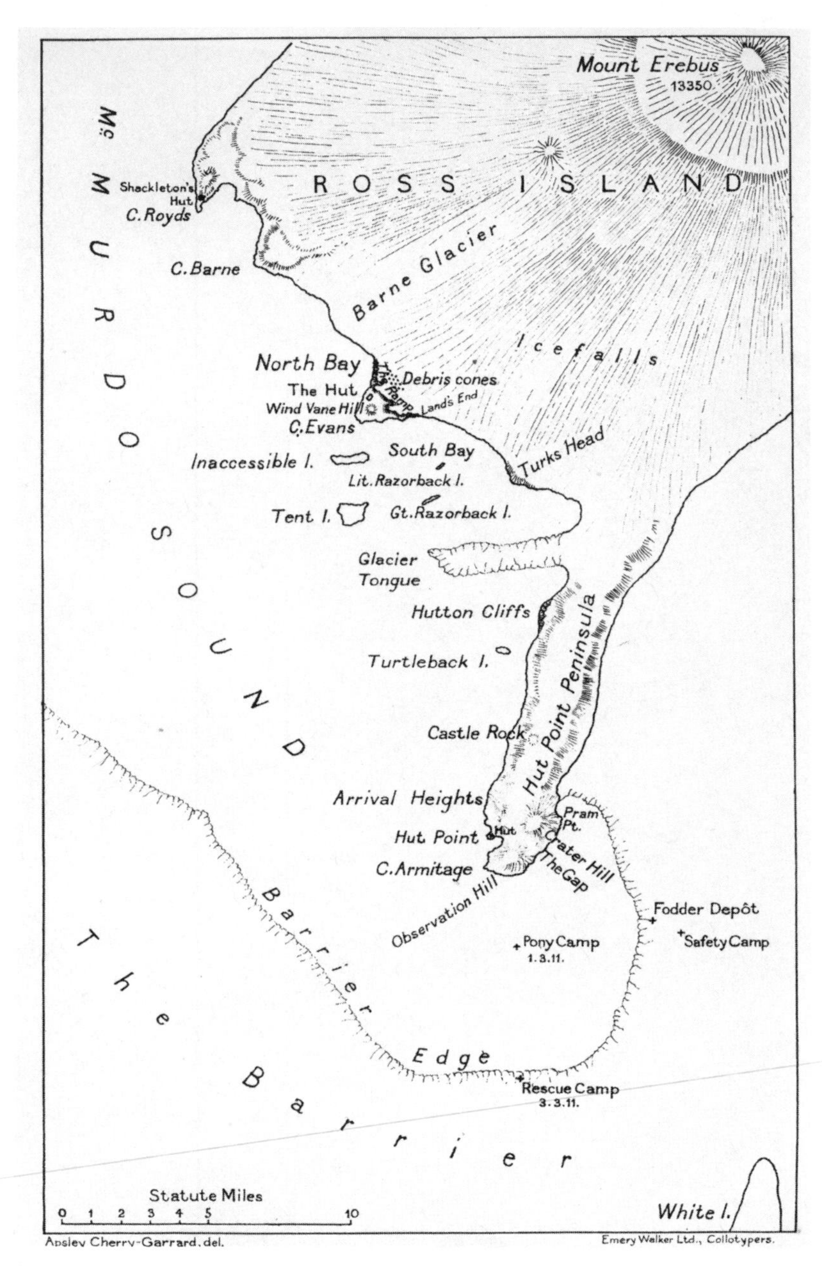

Ross Island map showing winter quarters for the *Discovery*, *Nimrod*, and *Terra Nova* expeditions, from Apsley Cherry-Garrard's *Worst Journey in the World* (New York, 1922).

AN EMPIRE OF ICE

Also by Edward J. Larson

A Magnificent Catastrophe: The Tumultuous Election of 1800, America's First Presidential Campaign (2007)

The Creation-Evolution Debate: Historical Perspectives (2007)

The Constitutional Convention: A Narrative History from the Notes of James Madison (with Michael Winship) (2005)

Evolution: The Remarkable History of a Scientific Theory (2004)

Evolution's Workshop: God and Science on the Galapagos Islands (2001)

Summer for the Gods: The Scopes Trial and America's Continuing Debate Over Science and Religion (1997)

Sex, Race, and Science: Eugenics in the Deep South (1995)

Trial and Error: The American Controversy Over Creation and Evolution (1985)

AN EMPIRE OF ICE

Scott, Shackleton, and the Heroic Age of Antarctic Science

EDWARD J. LARSON

Yale
UNIVERSITY PRESS
New Haven & London

Plates 5 and 23 courtesy of Scott Polar Research Institute Archives. Maps, figures, and plates from Robert Scott, *The Voyage of the "Discovery"* (New York, 1905), Roald Amundsen, *The South Pole* (New York, 1922), and James Ross, *A Voyage of Discovery and Research in the Southern and Antarctic Regions* (London, 1847), courtesy of the University of Georgia Libraries. Maps, figures, and plates from Ernest Shackleton, *The Heart of the Antarctic* (Philadelphia, 1909), and Apsley Cherry-Gerrard, *The Worst Journey in the World* (New York, 1922), courtesy of UCLA Libraries. Maps, figures, and plates from R. F. Scott, *Scott's Last Expedition* (New York, 1913), courtesy of Pepperdine University Libraries.

Yale University Press books may be purchased in quantity for educational, business, or promotional use. For information, please e-mail sales.press@yale.edu (U.S. office) or sales@yaleup.co.uk (U.K. office).

Designed by James J. Johnson and set in Monotype Dante type by Duke & Company, Devon, Pennsylvania. Printed in the United States of America by Thomson-Shore, Dexter, Michigan.

The Library of Congress has cataloged the hardcover edition as follows:

Larson, Edward J. (Edward John)
An empire of ice : Scott, Shackleton, and the heroic age of Antarctic science / Edward J. Larson.
p. cm.
Includes bibliographical references and index.

ISBN 978-0-300-15408-5 (clothbound : alk. paper) 1. Antarctica—Discovery and exploration—British. 2. Scientific expeditions—Antarctica—History—20th century. 3. Scott, Robert Falcon, 1868–1912. 4. Shackleton, Ernest Henry, Sir, 1874–1922. I. Title.
G872.B8L37 2011
919.8′9—dc22
2010044396
ISBN 978-0-300-18821-9 (pbk.)

A catalogue record for this book is available from the British Library.

10 9 8 7 6 5 4 3 2 1

IN THE SPIRIT OF

Robert Falcon Scott,

who wrote to his wife from his final camp on the Polar Sledge Journey

about their young son,

"Make the boy interested in natural history, if you can";

I dedicate this book to our son,

Luke Anders Larson,

who has expressed interest in all things Antarctic

CONTENTS

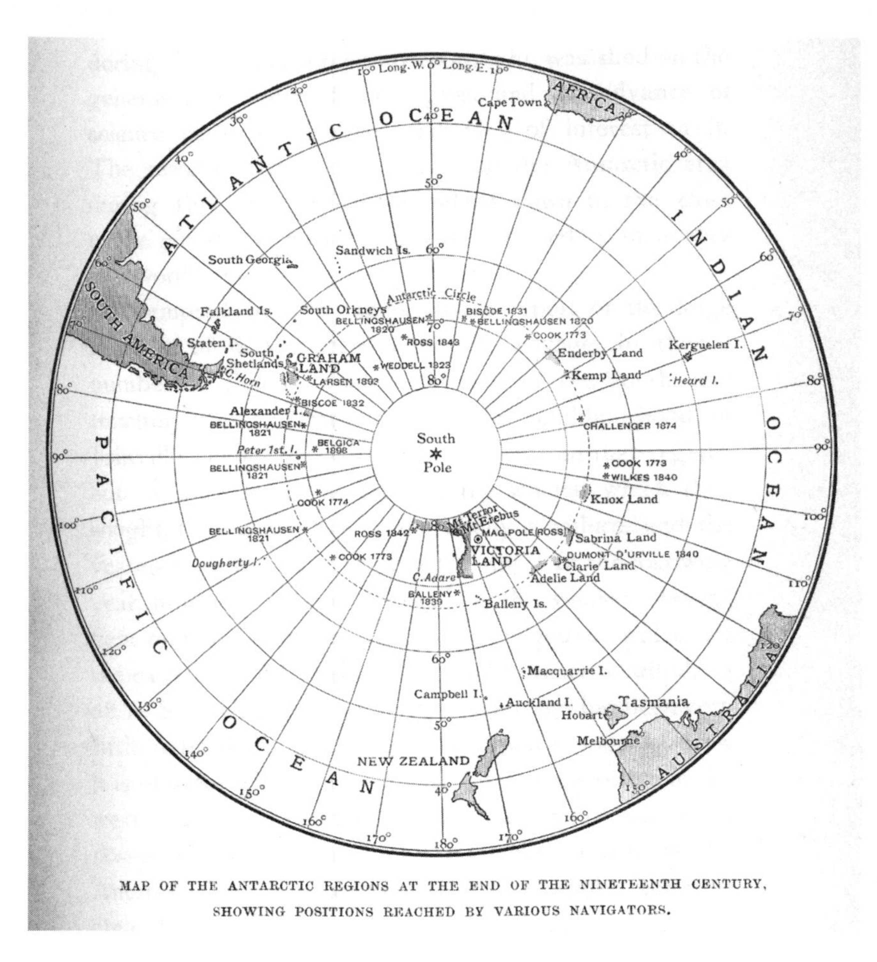

MAP OF THE ANTARCTIC REGIONS AT THE END OF THE NINETEENTH CENTURY, SHOWING POSITIONS REACHED BY VARIOUS NAVIGATORS.

Antarctic map showing what was known in 1900, before the *Discovery* expedition, including coastlines and the dates and places of prior expeditions, from Robert Scott's *Voyage of the "Discovery"* (New York, 1905).

PREFACE

When I tell friends that I'm writing a book about the Heroic Age of Antarctic exploration, they typically respond in one of two ways. Some say how much they admire Ernest Shackleton's leadership style, while others question Robert Scott's tactics in trying to reach the South Pole first. Both responses are telling. A century after their exploits, these two men are still widely known for their personal achievements, but their fame rests largely on how they dealt with adversity in their efforts to reach the geographical South Pole. That, most people assume, is why they went to Antarctica; much else about their expeditions is forgotten.

This book is neither a paean to Shackleton's leadership nor a critique of Scott's choices. It is about what was central to British efforts in the Antarctic. In the era before World War I, when Antarctic exploration was largely a British project, that project was largely concerned with science.

Scott led two expeditions to Antarctica during the first twelve years of the twentieth century; Shackleton led one. Scott's first was part of an international program also involving German and Swedish teams to explore the Antarctic that was fundamentally scientific in design and execution, although of course it had military, commercial, ideological, and personal motives as well. The ensuing expeditions by

Shackleton and Scott followed directly on Scott's first effort and adopted its basic scheme. All three British expeditions entered Antarctica through the Ross Sea. They form a logical unit that stand apart from other expeditions of the so-called Heroic Age, including Shackleton's second, both in their organization and in their impact.

If the race to the South Pole eventually consumed Scott, it was never at the expense of science. His two expeditions and Shackleton's 1907–9 venture carried enormous scientific baggage. If getting to the pole first was Scott's overriding objective, he went about it the wrong way. If he meant to get to the pole first while doing meaningful science along the way, he did it right—but in doing so, he fatally handicapped himself in a contest against Roald Amundsen, a polar adventurer of proven ability who cared only about winning the race. Focus empowered him. Scott and Shackleton served many masters, one of which was the British conception of scientific discovery, exploration, and conquest.

Any account of the three British Antarctic expeditions between 1901 and 1913 inevitably touches on Shackleton's leadership, Scott's choices, and the race to the pole. But these expeditions were complex enterprises. Science wove through every part of them, both influencing and being influenced by their other aspects—including such critical intangibles as leadership and choices. I know of no better way to understand the whole of these expeditions than through the lens of their research. Fortunately, the less-told tale of the explorers' scientific activities is often as gripping as the story of their polar quest.

Examining the astounding research efforts of these expeditions also illumines the fundamental place of science in Victorian and Edwardian British culture. Britain built and sustained its global empire during this period. In doing so, explorers and imperial officers took Western science to the four corners of the world—measuring, mapping, and collecting specimens as part of their program to subdue alien territory and make it British. The proud citizen of a nation that had recently cast off foreign rule, Amundsen came from a different tradition than Scott and Shackleton, and had different goals. Empire is not only about the physical conquest of territories; for the British, it was always about scientifically exploring and systematically exploiting them even as the definition and conception of science itself evolved. In a sense, then, this story is not only about the explorers' science in its various and contested forms. It is also about power and politics; culture and commerce; hubris and heroism at the end of the Earth.

Books about the Heroic Age of Antarctic exploration could fill a library. They fill a bookcase in mine. Yet except for many of the participants' published diaries and memoirs, most of these books—including some of the best—say little about science. There are notable exceptions. David Yelverton's thorough account of Scott's first expedition, *Antarctica Unveiled,* fully incorporates research activities into the overall account. T. H. Baughman does so too in his works on the dawn of the Heroic Age. Modern-day Antarctic researcher Susan Solomon included science in her retelling of Scott's second expedition, *The Coldest March.* Although focused on later periods, G. E. Fogg's technical *A History of Antarctic Science* covers the early expeditions in some detail. In his 1967 book, *South to the Pole,* L. B. Quartermain provides a comprehensive account of expeditions to the Ross Sea region though the Heroic Age. There are others. Here, I attempt to place the research work of three well-known expeditions into a broad scientific, cultural, and social context reaching back into the Victorian era and across to other expeditions of the period. The coverage may be less than encyclopedic at times, but everywhere it is more than representative.

With numerous scientific disciplines and multiple expeditions coming and going, narrative structure became challenging. A chronological account would leave the reader bouncing between a dozen fields of science without context or closure. Instead, I have layered the narrative by major disciplines: biology, geography, geology, glaciology, meteorology, oceanography, paleontology, and terrestrial magnetism. This structure means that each chapter starts anew, in the nineteenth century or before, and carries the story of its particular science into the early twentieth century or beyond.

The chapters are organized to minimize repetition and give some sense of the unity of the whole, yet that unity was more fundamental than this approach suggests. Oceanography, marine biology, and meteorology so merged in the ecological field studies of these expeditions that I combined them into a single chapter. Geology, glaciology, and physical geography are separated into different chapters largely because of the vast amount of new findings generated by these expeditions, but in fact those findings drew these fields closer together. By exposing the stark Antarctic environment to scientific analysis for the first time, these expeditions helped to reveal nature's fundamental unity to discipline-divided scientists and thus lay a foundation for modern concepts of ecology. Scott and Shackleton loom large in some of the resulting chapters; in others they are eclipsed by such lesser-known figures as Louis Bernacchi, Edgeworth David, Frank Debenham, Hartley Ferrar, Douglas Mawson, James Murray, Raymond Priestley, Griffith Taylor, Edward Wilson, and Charles Wright. On these expeditions, many people played influential roles.

Science involves measurement. For the Antarctic explorers, this mostly meant length, depth, height, weight, area, volume, and temperature. They generally used imperial units: inches, feet, miles, fathoms, leagues, pounds, tons, acres, gallons, and degrees Fahrenheit. So far as possible, this book follows their conventions. Compounding the confusion for modern readers, the explorers used three types of miles—statute, nautical, and geographical—often without differentiation, and sometimes used a short ton instead of the British long ton. In accord with current American usage, unless otherwise noted I have attempted to convert their figures to statute miles and short tons.

Early-twentieth-century explorers called some prominent geographical features by names that are no longer widely used. To avoid confusion, I have generally followed the explorers' conventions. For example, they typically referred to the entire Ross Ice Shelf as the Great Ice Barrier, a term now generally reserved to the shelf's seaward edge, and hailed the southern continent's defining feature, the East Antarctic Ice Sheet, as simply the Polar Plateau. Further, in excerpts drawn from the records of early explorers, McMurdo Sound is often called a bay.

Conducting scientific research in Antarctica has always required collaboration, and this is true for my study of its history as well. This book is the direct product of my participation in the National Science Foundation's 2003–4 Antarctic Artists and Writers Program. Always traveling with others, and frequently in the company of experts, through this program I saw much of what the early explorers saw, from Ross Island and the Great Ice Barrier to Beardmore Glacier and the South Pole. On December 18, 2003, exactly one hundred years after Scott, Edgar Evans, and William Lashly became the first humans to enter an Antarctic dry valley, I retraced their steps through Taylor Valley with the longtime manager of its research camp, Rae Spain. A few weeks later, I camped near Shackleton's winter quarters at Cape Royds with David Ainley, who has studied the cape's Adelie penguins for years. Both Spain and Ainley know the region's human history. Such experiences made this book possible. Having taken this extended trip during my tenure as chair of the University of Georgia's History Department, I want to thank Vici Payne for keeping the office going in my absence.

Many of the explorers' scientific papers, field notes, diaries, and letters are published; many unpublished ones are held in public archives. For access to the unpublished sources, I wish to acknowledge the archives and thank the archivists and librarians at Scott Polar Research Institute in Cambridge, the Royal Geographical Society of London, the International Antarctic Centre and Canterbury Museum at Christchurch, New Zealand, the Royal Society in London, the National Maritime Museum of Greenwich, the South African Astronomical Observatory in Cape Town, Byrd Polar Research Center at The Ohio State University, and the McMurdo and Amundsen-Scott South Pole Stations in Antarctica. For published resources from their own collections and interlibrary loan sources, I received particular help from the UCLA, University of Georgia, and Pepperdine University libraries and reference librarians, especially from Pepperdine's Jodi Kruger and Marc Vinyard.

I can only begin to identify all other individuals who have contributed to this book. William Frucht, my editor at Yale University Press, time and again kept this effort on track and brought it to completion, just as he did my earlier book about the history of science on the Galápagos Islands, *Evolution's Workshop*. I cherish his continuing support. Through my association with the Antarctic Artists and Writers Program, many polar administrators, educators, and scientists have offered advice or information, including National Science Foundation program manager Vladimir Papitashvili, Umram Inan, Evans Paschal, Serap Tilav, Jöerg Hörandel, Collin Blaise, Eyal Gerecht, Antony Stark, Nicolas Tothill, Thomas Nikola, William Holzaphfel, Robert Garrott, Laurie Connell, Paul Ponganis, Carol Landis, Diane McKnight, Justin Joslin, Alexandre Tsapin, Ross Virginia, Michael Poage, Diana Wall, Ralph Harvey, Robert Smalley, Bruce Luyendyk, Luann Becker, Howard Conway, Fred Eisele, Detlev Helmig, Stephen Warren, Stephen Hudson, Glen Kinoshita, Alan Campbell, and (above all) Antarctic Artist and Writers Program manager Guy Guthridge. If I had not dedicated this book to my son, I would have dedicated it to Guy. Most of all, my thanks go to my wife, Lucy, and our children, Sarah and Luke. During my work on this book, they have endured my absences and preoccupations without displaying any bitterness toward penguins.

AN EMPIRE OF ICE

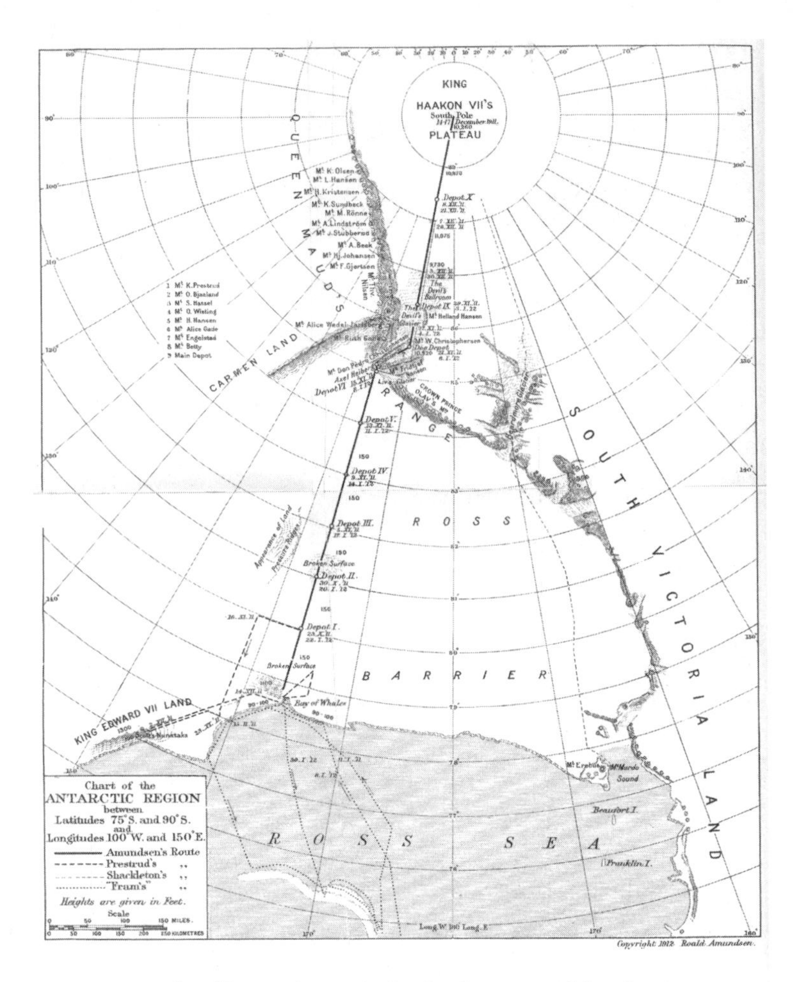

Route of Roald Amundsen to the South Pole in 1911, with line showing Ernest Shackleton's 1908–9 route, which was roughly followed by Scott in 1911–12, from Amundsen's *South Pole* (New York, 1913).

CHAPTER 1

"Three Cheers for the Dogs"

HE STOOD IN TRIUMPH AND TREPIDATION. IT WAS the evening of November 15, 1912. A proud, plain-speaking Norwegian adventurer, Roald Amundsen rose to address a packed house at London's elite Royal Geographical Society after having bested better-equipped and better-funded British explorers in attaining a long-prized goal. He had reason to tremble. Some in the audience saw him as a jackal in a den of lions.

His talk would be modest, focused more on technical details of the journey than on the end accomplishment—but it could not be modest enough to please many of his British listeners. They, in turn, could not avoid insulting him even had they wanted to do so. For the second time in his life, he had achieved what Britain's greatest heroes could not, but he had done it in a way that they disdained to attempt.

In 1912, when Amundsen made this second triumphal appearance before the society, London reigned over the most extensive empire in the history of the world. For three centuries, British explorers had led Europe in the discovery of other lands and seas. The Royal Geographical Society, or RGS, traced its origins to 1788, as the Association for Promoting the Discovery of the Interior Parts of Africa, and had succeeded famously in its original goal through its support of David Livingstone,

Richard Burton, John Speke, Henry Stanley, and other renowned explorers. Under the patronage of Queen Victoria, it extended its reach to the ends of the earth. Antarctica, the last large blank space on world maps, had by 1900 become a focus of its ambitions. The South Pole took on aspects of a holy grail.

British geographers of the late Victorian and Edwardian era viewed themselves as scientists and their expeditions as grand enterprises of science. Simply reaching the head of the Nile, the high Himalayas, or the South Pole was not enough. An RGS explorer had to conduct research along the way. A series of RGS-endorsed expeditions had been opening the way to the pole for more than a decade when, late in 1911, Amundsen stole a march on a team already in the field to capture the prize by questionable means. Hailed for this achievement throughout most of the Western world, Amundsen was all but required to address the leaders of British geographical science and receive their validation of his effort. He did not want to come but could scarcely decline their summons. The Royal Geographical Society's status as the arbiter of world geography was well earned.

The British boasted a long history of exploration and claimed a certain province over the far south, where early geographers thought a large landmass must exist to counterbalance the continents of the north. During the 1770s, the Admiralty launched a scientific expedition under the command of James Cook to look for this hypothesized southern land. "January 17th, 1773, was an epoch in the world's history," RGS Librarian Hugh Robert Mill declared in 1905, "for just before noon on that day the Antarctic circle was first crossed by human beings." The intrepid Cook crossed the Antarctic Circle twice more on the same voyage but retreated each time before dense pack ice without sighting land. By circumnavigating the globe at roughly latitude 60° south, he established that, if an Antarctic continent existed, it must lie in the far south behind a daunting blockade of sea ice. "I will not say it was impossible any where to get farther to the South; but attempting it would have been a dangerous and rash enterprise," Cook wrote in his journal. "It was, indeed, *my* opinion, as well as the opinion of most on board, that this ice extended quite to the pole, or perhaps joined to some land, to which it had been fixed from the earliest time."[1]

Later British explorers thought otherwise. In 1839 the Admiralty commissioned a second expedition to Antarctic waters. James Clark Ross, already famous as the first European to reach the wandering North Magnetic Pole, was given two sturdy wooden ships, HMS *Erebus* and *Terror,* and a charge to make magnetic observations throughout the deep southern seas. By this time, sealers, whalers, and expeditions from various countries had probed the edges of the ice pack and returned with reports of isolated bits of land.

"Impressed with the feeling that England had ever *led* the way of discovery in the southern as well as in the northern regions," Ross commented, "I considered it would have been inconsistent with the pre-eminence she has ever maintained, if we were to follow in the footsteps of the expedition of any other nation." Instead he plowed through the ice pack south of New Zealand and found a vast open sea with a mountainous western coast that he named Victoria Land for his young queen. "It was an epoch in the history of discovery," the RGS's Mill later wrote, "the magic wall from before which every previous explorer had to turn back in despair, had fallen into fragments at the first determined effort to break through it."[2]

Sailing south along the Victoria Land coast in the sea later named for him, Ross encountered at about latitude 78° south what he described as "a perpendicular cliff of ice between one hundred and fifty and two hundred feet above the level of the sea, perfectly flat and level on top, and without any fissures or promontories on its even seaward side."[3] The awestruck captain found that this "Great Ice Barrier" extended eastward from the Victoria Land coast for hundreds of miles.

He realized it would prevent anyone from sailing farther south. Despite this obstacle in the way to the pole, Ross had found an exposed coastline with majestic mountains and, jutting from the Ice Barrier across the Ross Sea's McMurdo Sound, a large island that was later named for him. He never set foot on the Antarctic mainland, but his namesake island became the base for many later efforts to probe the southern continent.

In 1901, after years of prodding by its president, Clements R. Markham, the RGS cosponsored the first British land expedition to the southern continent. Aboard the purpose-built wooden ship *Discovery*, which wintered over for two years at Ross Island's Hut Point with a select team of scientists, officers, and sailors under Royal Navy commander Robert Falcon Scott, the British National Antarctic Expedition became the first to send parties south across the Ice Barrier and west over the Victoria Land mountains. A team consisting of Scott, Ernest Shackleton, and Edward Wilson set a new farthest south record on the last day of 1902 before turning back at the extreme end of their endurance, at just over latitude 82° south. They had covered almost five hundred miles on foot with heavy sledges. "Whilst one cannot help a deep sense of disappointment in reflecting on the 'might have been' had our team remained in good health," Scott wrote in his published journal, "one cannot but remember that even as it is we have made a greater advance towards a pole of the earth than has ever yet been achieved by a sledge party."[4]

Shackleton returned to Antarctica five years later leading a privately funded expedition aboard the forty-year-old converted sealer *Nimrod*. Accompanied by a small land party that included several scientists, he wintered at Cape Royds on Ross Island before heading south with three men across the Ice Barrier, up a glacial pass through the mountains of South Victoria Land, and onto the vast Polar Plateau. They man-hauled their sledge to within 120 miles of the pole before being forced to turn back or face certain death by starvation. "We have shot our bolt, and the tale is latitude 88° 23′ South, longitude 162° East," Shackleton wrote on January 9, 1909. "We hoisted her Majesty's flag and the other Union Jack afterwards, and took possession of the plateau in the name of his Majesty," King Edward VII.[5]

Scott, sailing from England aboard *Terra Nova,* with his sights locked on the South Pole, had his second Antarctic expedition under way before Amundsen's ship, *Fram,* departed on September 9, 1909, with the same ultimate destination. Scott brought along more scientists than any prior Antarctic expedition; Amundsen took none. The two teams spent the Antarctic winter at Ross Sea harbors four hundred miles apart, with the British base at Ross Island's Cape Evans and the Norwegian one at a cleft in the Ice Barrier known as the Bay of Whales. They set off with sledges for the pole within days of each other—Amundsen on October 20, 1911, and Scott on November 1. Many Britons viewed the entire Ross Sea basin as their domain by right of discovery and prior exploration. The Norwegians were virtual trespassers.

For all the attraction of the South Pole, the Arctic held greater fascination for the British during the nineteenth century—and here too Amundsen had come late to the game. British interests in the Arctic regions of North America began with the practical purpose of finding a Northwest Passage for sea trade with Asia. In 1497, soon after Christopher Columbus returned from his epic first voyage to the New World, King Henry VII of England sent John Cabot in search of a northern route around the Americas. He found none. A succession of expeditions over the next three centuries, while cementing British claims to the Hudson Bay region under Western concepts of acquisition by European discovery, established that if a Northwest Passage existed through Canada's Arctic Archipelago, then it was likely blocked by ice most of the year. Still, hopes of finding open water at the top of the world persisted into the mid-1800s.

Idle years for the British navy following the defeat of Napoleon in 1815 led Admiralty Second Secretary John Barrow to promote naval expeditions to the Canadian Arctic as a means to engage sailors and officers during peacetime, expand the empire, and make scientific and geographical discoveries. Surplus British warships were soon probing the far northern seas and coasts under the command of such veteran naval officers as John Ross, William Parry, and John Franklin—all of whom had served with distinction in the Napoleonic Wars. Ross's nephew James Clark Ross, the future Antarctic hero, got his education in polar exploration by participating in six of these Arctic voyages from 1818 to 1833 under the tutelage of his uncle and Parry. Coinciding with the Romantic movement in the arts, these expeditions provided grist for countless books, paintings, and poems, including Mary Shelley's groundbreaking 1818 science fiction tale, *Frankenstein*. Shelley's tragic hero pursues his monstrous creation to the frozen north, where they encounter an icebound British Arctic expedition, which carries back their story of scientific hubris, death, and self-immolation on the polar ice. "I shall quit your vessel on the ice-raft which brought me hither," the monster tells the expedition's leader, "and shall seek the most northern extremity of the globe; I shall collect my funeral pile and consume to ashes this miserable frame."[6] The North Pole already had become an ultimate destination.

Leaders of actual British Arctic expeditions returned with their own tales of life, death, and science in the far north, which they often related in popular articles and books. Franklin became famous as "the man who ate his boots" after surviving a harrowing overland expedition in which most of his men starved to death and the rest resorted to eating lichen, shoe leather, and (by some accounts) their fallen comrades. For successful officers—especially those who published gripping accounts of their exploits—Arctic service provided a means to attain promotion in a peacetime navy, as well as celebrity status and entry into elite social circles. Following their initial triumphs,

Parry married into the aristocracy while Franklin wed the poet Eleanor Porden and, after her death, the wealthy world traveler Jane Griffin.

By 1845, parts of the Arctic Archipelago had been surveyed from either the Atlantic or Pacific end, but no one had completed a voyage through it. At age fifty-nine, after an interlude as a colonial governor, Franklin accepted command of Ross's fabled *Erebus* and *Terror* to complete the passage in the course of taking magnetic readings around the North Magnetic Pole, but these ships became trapped in the ice and never returned. Assuming that the explorers would have abandoned their ships and proceeded on foot, the Admiralty dispatched a series of land and sea expeditions to find them. When these failed, Franklin's wife sponsored four expeditions of her own and offered a reward that spurred on others before conclusive evidence showed that Franklin and his men had died either on board the icebound ships or during their attempted trek to safety. Inuit accounts of cannibalism among some starving crewmembers, at first discounted but later proven, darkened these reports.

To search for Franklin on the ice and land, the Royal Navy refined techniques of man-hauling heavy sledges. Although participants reportedly described this as "about the most severe work to which man has ever been put, at least in modern times," it served as an appropriate means of winter transport in the far north for young sailors disciplined for teamwork and accustomed to handling ropes. The native people would have used dogsleds, but these required training that the searchers lacked. After a disastrous Arctic expedition during 1875–76, one former sledger wrote to the niece and companion of Franklin's widow about continuing the brutal practice, "I would confine every one who proposed such a thing in a Lunatic Asylum, burn every sledge in existence and destroy the patterns."[7] He did not reckon with the force of navy tradition.

Through a dozen publicly or privately funded expeditions and over a hundred sledge trips, during which more sailors and ships were lost than on the original voyage, the Franklin searches greatly extended the survey of the Canadian Arctic. Yet no single ship traversed the entire Northwest Passage. That distinction was left to a small, shallow-draft fishing sloop, *Gjoa,* commanded by Amundsen with a crew of six. Inspired by Franklin's 1824 book about searching for the Northwest Passage and by the outpouring of nationalistic euphoria after six Norwegians led by Fridtjof Nansen had skied across Greenland in 1888, Amundsen in his teens resolved to become a polar explorer. "Strangely enough," he later wrote about Franklin's book, "the thing in Sir John's narrative that appealed to me most strongly was the sufferings he and his men endured. A strange ambition burned within me to endure those same sufferings." Of Nansen, he added, "The 30th May, 1889, was a red-letter day in many a Scandinavian boy's life. Certainly it was in mine. That was the day when Fridtjof Nansen returned from his Greenland Expedition."[8] Amundsen's remote but revered seafaring father had died three years earlier; in Nansen, he found a hero and mentor.

After his mother's death freed him from her demands that he become a land-bound physician, Amundsen openly pursued his polar dreams. Descended from a family of ship owners and captains, in 1897, the twenty-five-year-old threw in his lot with a barebones Belgian expedition to Antarctica that became the first to winter at the southern continent when its ship, *Belgica,* became trapped in the sea ice west of the Antarctic Peninsula. "For thirteen months, we lay caught in the vise of this ice field," Amundsen recalled. "Two of the sailors went insane. Every member of the ship's company was afflicted with scurvy, and all but three of us were prostrated by it."[9] In his memoirs, Amundsen credited himself and the expedition's American doctor, Frederick Cook, with saving the expedition by directing the crew to eat fresh seal meat and to cut a channel in the ice from their ship to a nearby melted basin that eventually opened to the sea.

Having earned his spurs in polar exploration and gained Nansen's backing, Amundsen organized his ambitious cruise through the Northwest Passage, which lasted from 1903 to 1906. Nansen was by this time a world-renowned professor of zoology and oceanography. To win his support, Amundsen cast this expedition as primarily a scientific effort to relocate and study the movement of the North Magnetic Pole. "He emphasized that this investigation of the magnetic pole was the expedition's mission statement, the scientific core which gave it legitimacy," Nansen later recalled, "and that, as they were already there, they might as well include the Northwest Passage." To prepare, Amundsen briefly studied in Germany under Georg von Neumayer, a leading expert in terrestrial magnetism and proponent of polar exploration. He also secured the RGS's endorsement for the effort. "My expedition must have a scientific purpose as well as the purpose of exploration," he noted. "Otherwise I should not be taken seriously and would not get backing."[10] For Amundsen, however, the tail wagged the dog.

As the expedition proceeded, Amundsen increasingly turned over the scientific work to a young assistant, Gustav Wiik. "Wiik works continually on the magnetic north," one crewman noted during the second winter. "The Governor [Amundsen] and the Lieutenant [Godfred Hansen] read novels and smoke and go for walks from time to time. It is unbelievable that a man can change like the Governor has in the course of one year. Last year he worked constantly with his observations. This year he has done nothing."[11] The crewman exaggerated. During two long winters frozen in place, Amundsen did learn valuable polar survival and dogsledding skills from the local Inuit people, with whom he freely bartered for food, animal-skin clothing, and women. Most important, with Amundsen in command, *Gjoa* made it through a passage that had blocked every earlier ship that attempted it.

Word of the successful transit of the Northwest Passage, telegraphed from Alaska after two years without communication, touched off revelry in Norway, which had secured its independence from Sweden during Amundsen's absence. For many Norwegians at home and abroad, Amundsen became a second Nansen. "I attend one celebration after the other and have no time for much else," Amundsen wrote of his return journey by train across the United States, which featured paid lectures at cities with large Norwegian populations. A European tour followed, which included an appearance before the RGS in London on February 11, 1907. Then came the obligatory book, *The Northwest Passage,* and book tour. "Don't forget that from now on you must consider yourself a businessman," Amundsen's brother turned business manager, Leon, telegraphed to the explorer soon after his arrival in Alaska. "You might make a lot of money on a lecture tour of America, which is of course what you want."[12]

Amundsen received honors wherever he went, including Norway's highest decoration, but less so in England, where there was a sense of diminished achievement. He had taken a small motored sloop—some called it a yacht—on a route made known through decades of British sacrifice on tall ships and with heavy sledges. Nansen's English agent, when asked about the prospects of Amundsen's lecturing in Britain, advised him, "Your spectacular expedition attracted attention among the scientific public, but has not caught the imagination of the general public sufficiently to make the lecture tour a financial success." The London press virtually ignored the Norwegian's feat, and the British government, which once promised a monetary award to the first person who found a northern sea route to the Pacific, declined to pay him. The reward, he was told, had already had been disbursed among the Franklin searchers who, the British maintained, had virtually accomplished it. Some sentimentalists in Britain even credited Franklin's much-romanticized lost expedition with achieving the feat. "However much these able men deserved remuneratory rewards for their hardships and achievements," Amundsen complained about the distribution of credit and payments, "the voyage of the *Gjoa* stands as the first and only actual navigation of the Northwest Passage."[13]

"There is no doubt that if you had returned home via Cape Horn with your ship," RGS Secretary John Scott Keltie explained to Amundsen, "it would have made a big impression on the British public and thereby you could possibly have got more money from papers and publishers."[14] But Amundsen was too busy capitalizing on the past expedition and preparing for his next one to sail *Gjoa* home. It remained on display in San Francisco until 1972, before it finally returned to Norway.

The RGS was Amundsen's toughest audience. Reviewing Amundsen's book for the RGS *Journal,* the society's immediate past president and longtime champion of polar exploration, Clements Markham, expounded on prior British expeditions to the region and noted Amundsen's good fortune at finding the passage ice-free, which Markham depicted as a chance event. Scott Keltie even suggested that the revered Nansen, who was then serving as Norway's ambassador in London, deliver Amundsen's address so that it could be better understood. Amundsen gave his own talk, which the society heard at its clubhouse, having declined to book a grand hall such as the Royal Albert, where it had received Nansen and would receive Shackleton after their polar triumphs. In his formal introduction of Amundsen's address, RGS President George T. Goldie, a former British administrator in Africa, underscored the expedition's "very small" size and ended by noting its leader's low-key reception in England. "I wonder what would have been the effect a century ago if it had been announced that some one was going to address a meeting describing his voyage through the North-West Passage?" he asked. "I do not think the Albert Hall would have sufficed if it had existed in those days. I now call upon Captain Amundsen to read his paper."[15]

Amundsen tailored his address to his audience, which included aging veterans of the Franklin searches. "To Sir John Franklin must be given the honor of having discovered the North-West Passage," he began. He attributed his success in being the first to sail through it to "the rich fund of experience gained by English navigators in those regions." He then devoted virtually his entire address to the expedition's magnetic research and his anthropological observations of native peoples. In short, he gave a science lecture. Amundsen said so little about the actual passage from known regions in the east to known ones in the west that, if anything, he made it sound too easy. Listeners could take his remarks as a backhanded rebuke of British failures.[16]

Nansen followed Amundsen to the dais. Ever the diplomat—he later received the Nobel Peace Prize for his work with the League of Nations—Nansen noted how this "lucky man" had used the discoveries of prior British explorers to make his crossing of the Northwest Passage. "It shows us a good example of the way in which British and Norse sailors work together," the ambassador noted. "I think we may say we belong to the same race." Nansen also dwelt on Amundsen's magnetic research. "He could have done the North-West Passage long before he did," Nansen said, "instead of making scientific observations, which, I am afraid, very few of the public and of you appreciate as they ought to be appreciated." Perhaps expressing his hopes rather than his beliefs, Nansen concluded that his protégé "knows what is of importance and what is not; he knows not to do sensational things when he has good work to do, but he can appreciate sensational things at the same time, as he has shown us. And we may see him start again on a new exploration, and I feel certain, next to his own country, he will have many well-wishers in this country."[17] By that expedition's end, however, some in Britain would come to see him in a dark light in which a sensational dash to the pole eclipsed the important work of science.

According to his best biographer, Tor Bomann-Larsen, Amundsen never forgot the perceived slights he received from the British following the *Gjoa* expedition. Later in 1907, invited back to receive the RGS's Patron's Medal for "magnetic research in the region of the north magnetic pole," rather than for transiting the Northwest Passage, Amundsen never even replied.[18] A bold new adventure loomed ahead.

In 1907, a polar explorer with Amundsen's ambition could savor two epic unclaimed destinations: the two poles. Science beckoned south, where the pole potentially sat on a high plateau of an unknown continent. Although scientists remained interested in the Arctic basin, the North Pole was merely a mathematical point on a vast expanse of sea ice. The science-minded British appeared intent on reaching the South Pole: Shackleton was preparing to leave on *Nimrod,* and Scott, not long back from his first try, was already negotiating with the RGS for another shot at it. Britain had all but abandoned the North Pole to the Americans, who seemed to care only about setting records. In a series of well-publicized efforts over the preceding two decades, the American Robert E. Peary and his assistant, Matthew Henson, had used Inuit dogsledders to push ever farther north. Although some openly doubted his claim—there was no way to confirm it—Peary reported having reached slightly over latitude 87° north in 1906 before being forced back. Calculating and obsessed, he would surely try again for the North Pole. Amundsen decided to contend for it.

A decade earlier, Nansen had set a prior "farthest north" by a means that incorporated science. Amundsen now prepared to emulate Nansen in order to reach his true goal: a pole.

Nansen had observed that the Arctic ice cap slowly circulates in response to underlying ocean currents. In an effort RGS leaders considered foolhardy, he secured Norwegian funding for a purpose-built ship, *Fram,* which he wedged into the sea ice north of Siberia with the intent of drifting across the pole in the circulating ice. The expedition could take years, during which time Nansen would study polar currents and climate as he drifted.

The plan worked, up to a point. The icebound *Fram* drifted north in a wide arc. When it became clear that the arc would fall short of the pole, Nansen and a colleague set off with skis, dogs, sleds, and kayaks for the pole. They established a new record of latitude 86° 14′ north—nearly 2° north of *Fram*—before turning back for a horrific sixteen-month journey home. Drifting with the circulating ice, the ship and its crew eventually broke free on the ice pack's far side, near Spitsbergen. This was the sort of adventure that inspired Amundsen.

While in London to deliver his RGS lecture on the *Gjoa* expedition, Amundsen apparently discussed his new scheme with Nansen. Nansen may have been alluding to it when he spoke of Amundsen's linking the scientific and the sensational in "a new exploration." By the end of 1907, Amundsen had Nansen's backing and loan of *Fram* for a second drift across the Arctic ice cap beginning in 1910—it would be only the third major voyage by the pride of Norway's fleet. In 1898, Otto Sverdrup had borrowed it for a scientific expedition to the Canadian Arctic.

To gain Nansen's support and government funding, Amundsen sold the expedition as an oceanographic and meteorological study of the Arctic basin, with the North Pole an implied but unstated bonus. "From the moment the vessel becomes fast in the ice, a series of observations will be begun, with which I hope to solve some of the hitherto unsolved mysteries," Amundsen told the RGS in January 1909, without mentioning the pole.[19] The society responded with a small grant—"and that's something," Amundsen commented privately. About the RGS secretary, Amundsen added, "Keltie is completely changed. He is the nicest, friendliest person you can imagine."[20]

To improve his chances of drifting near the pole, Amundsen decided to start much farther east than Nansen. "Our plan for the drift provided that we should enter the Arctic Ocean through the Bering Strait," he wrote. "Our route from Norway to Bering Strait was by way of Cape Horn."[21] This route around South America was fortuitous because when Amundsen's former *Belgica* colleague Frederick Cook claimed the North Pole in April 1908, and then Peary did so in April 1909, Amundsen was already scheduled to sail south. As the two Americans squabbled over who had reached the North Pole first and doubts arose over whether either actually did, Amundsen's established itinerary provided cover for the most radical possible change of plans.

Amundsen had little interest in reaching anywhere second—or third. He wanted firsts. Publicly, however, he argued the scientific merit of the Arctic drift expedition, assuring backers that it would proceed with or without the pole. He needed to keep up the appearance of a proper scientific expedition to retain his funding, and he stood little chance of gaining support for a South Pole gambit while Scott was preparing a massive expedition to the region. Confiding at first only in his brother, Leon, and without telling Nansen or the Norwegian government, Amundsen packed *Fram* for a dash to the South Pole. Ninety-seven sled dogs from Greenland were loaded on board—which made no sense if he could gather huskies in Alaska after the long voyage around the Horn—along with a prefabricated hut for wintering on the Ice Barrier. The crew never became suspicious, and the watchful Nansen remained oblivious despite having asked Amundsen "why on earth he'd got so many dogs."[22] They were not alone: nobody guessed the truth.

Fram's real destination was revealed only at its last port of call, the remote Portuguese island of Madeira. There, with Leon at his side, Amundsen informed members of his expedition and gave each of them a chance to disembark. No one did. Leon returned to Norway with letters to Nansen and the king, for simultaneous delivery, and a third for the Norwegian people. In them, Amundsen wrote that he still planned to complete the scientific drift across the Arctic but that finances forced him to detour south. "The masses" demand firsts,

Amundsen explained in his letter to Nansen, and money follows. Because the Norwegian government had refused to provide added funds for the drift expedition once Cook and Peary had claimed the North Pole, he lamented disingenuously, only "this extra excursion" could save "the expedition I originally intended."[23] By this time, Nansen could not interfere and Scott had already set sail. At his brother's direction, Leon wired a terse note to Scott in Australia: "Beg leave to inform you *Fram* proceeding Antarctic."[24]

"It was true that I had announced in my plan that the *Fram*'s third voyage would be in every way a scientific expedition, and would have nothing to do with record-breaking," Amundsen wrote two years later to justify his change of direction, "but in view of the altered circumstances, and the small prospect I now had of obtaining funds for my original plan, I considered it neither mean nor unfair to my supporters to strike a blow that would at once put the whole enterprise on its feet. . . . Scott's plan and equipment were so widely different from my own that I regarded the telegram that I sent him . . . rather as a mark of courtesy than a communication which might cause him to alter his programme in the slightest degree." Amundsen stressed a key distinction: "The British expedition was designed entirely for scientific research. The Pole was only a side-issue, whereas in my extended plan it was the main object. On this little détour science would have to look after itself."[25]

Following release of Amundsen's letter to the public, the press began to trumpet a "great international polar race"—much like the international car races of the period. If this was a race, it was between unequal competitors. Scott's *Terra Nova* carried sixty-six men, thirty-four of them in the shore party. They had various missions to perform, only one of which involved reaching the South Pole. Amundsen's *Fram* carried nineteen men, of whom only nine wintered in the Antarctic. All of them focused solely on getting a small party to the pole, though after the five-person polar party departed, three of those remaining made an exploring excursion east to the Antarctic mainland. Scott set his base on Ross Island, nearly seven hundred miles in a straight line from the pole. But his route called for going around obstacles and up glaciers with heavy sledges. Knowing this in advance, Amundsen established his base, called Framheim, at the Bay of Whales, which was more than sixty miles closer to the goal and, as it turned out, offered a more direct path to it.

After laying depots along their planned routes during their first Antarctic autumn and wintering in their respective bases, the two expeditions set off for the pole late in 1911. Amundsen started twelve days before Scott and traveled faster. "The plan was to leave the station as early in the spring as possible," the Norwegian explained. "If we had set out to capture this record, we must at any cost get there first."[26]

The race became increasingly uneven as it progressed. The Norwegians used expertly trained dogs to pull their sledges rapidly across the Ice Barrier, up the mountain passes, and over the Polar Plateau. Those not driving sledges used skis. The British struggled with experimental motorized tractors, white Manchurian ponies, and heavy, man-hauled sledges. Of course, only Amundsen knew in advance there would even be a race, and unlike Scott, he felt no compulsion to do science along the way. The RGS's self-anointed arbiters of Antarctic geography fumed. "What an imposter!" Markham exclaimed to Keltie in December 1910 about Amundsen on learning that the Norwegian was headed south. Although doubting that Amundsen could reach the pole, Markham conceded, "He is still a source of some anxiety."[27]

"At three in the afternoon a simultaneous, 'Halt!' rang out from the drivers," Amundsen wrote in his account for December 14, 1911. "They had carefully examined their sledge-meters, and they all showed the full distance—our Pole by reckoning. The goal was reached, our journey ended." The Norwegians saw no sign of Scott. They camped for four days and used every available means to confirm their position. After repeated observations of the sun, which circled steadily around them for twenty-four hours each day, they determined that the camp was some six miles off the mark and moved their tent accordingly. "On December 17 at noon we had completed our observations, and it is certain that we had done all that could be done. In order if possible to come a few inches nearer to the actual Pole, Hanssen and Bjaaland went out four geographical miles (seven kilometers) in the direction of the newly found meridian," Amundsen wrote. At dinner, they smoked cigars to celebrate.[28]

Scanning the horizon for Scott one last time, the Norwegians started home on December 17. They left a small tent with some excess provisions and a note for Scott, who they assumed would arrive shortly. "Welcome to 90°," read a leather strip sewed to the tent.[29] A Norwegian flag and a banner from *Fram* fluttered from a long pole extending from the tent's peak. Amundsen also deposited a letter to the Norwegian king, which he asked Scott to bring back as proof that both parties made it to the pole. Presumably neither wanted to repeat the bitter dispute that still clouded the discovery of the North Pole and tarnished its claimants' reputations.

"The going was splendid and all were in good spirits, so we went along at a great pace," reads the first sentence of Amundsen's chapter on the return trip to Framheim. The weather remained fair most of the way. Much of the route was downhill. "We always had the wind at our backs, with sunshine and warmth the whole time," Amundsen noted. As they skied down the long slope from the Polar Plateau, "the surface was absolutely polished, and for long stretches at a time we could push ourselves along with our sticks." Improvised sails were hoisted on the sledges. "The drivers stood so jauntily by the side of their sledges, letting themselves be carried over the plain at a phenomenal pace." At times the skiers struggled to keep up. Rations steadily increased for men and dogs as they passed their evenly spaced supply depots. "We could not manage more," Amundsen said of the food.

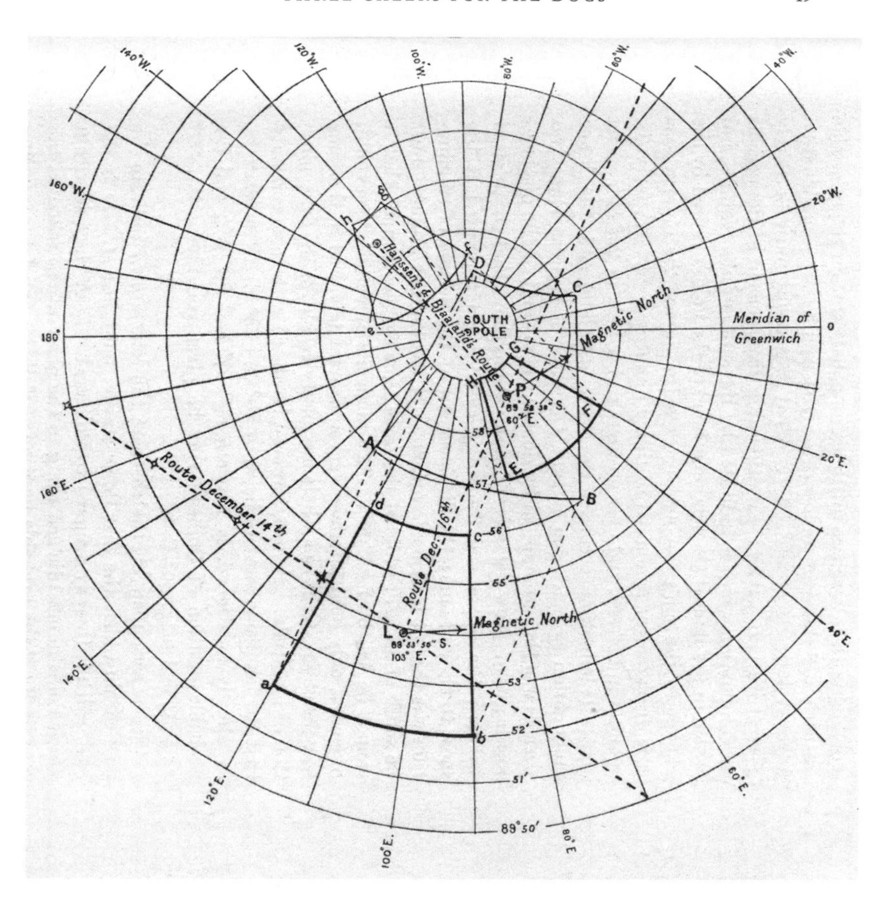

Sketch of the Norwegian expedition's route around the South Pole, December 15–17, 1911, with lettered polygons showing the field in which various observing stations must lie, from Roald Amundsen's *South Pole* (New York, 1913).

Some was left behind. "The dogs were bursting with health, and tugged at their harness." The polar party reached Framheim in good health and high spirits on January 25, 1912, after ninety-nine days out and back. Amundsen had gained weight. *Fram* had returned from winter quarters in South America. Six days later, the Norwegians sailed for Tasmania, where Amundsen telegraphed their success to the waiting world on March 8.[30]

The news made headlines throughout Europe, North America, and the British Empire, with the supposed race between Amundsen and Scott featuring prominently in the storyline. "The whole world has now been discovered," the *New York Times* declared on March 8.[31] Some outside Britain seemed to delight in a small band of Viking raiders tweaking mighty John Bull's nose. The *New York Times,* France's *La Main,* and the London *Daily Chronicle* paid Amundsen handsomely for exclusive rights to his first report. Other newspapers tried to scoop them or to run their own cobbled-together accounts, which led to a media frenzy. Celebrations erupted in Norway. At year's end, the *New York Times* ranked Amundsen's polar conquest as the number two news story of 1912, second only to the sinking of the *Titanic.*

Many in Britain could scarcely believe that a few Norwegians with dogsleds could beat their massive national expedition to the pole. When news of the *Fram*'s arrival in Tasmania leaked before Amundsen's carefully crafted first report to the *Daily Chronicle,* word spread through London that the Norwegians would say that Scott had beaten them to the pole. "Rumours began to-day that you had got to the Pole and reporters began to flock and telephone," Scott's wife, Kathleen, wistfully wrote to her husband in her diary on March 6, "but apparently there is no sort of foundation." After Amundsen squelched those rumors with his subsequent telegram, some Brits remained optimistic. "It still seems possible," an essayist wrote in the nationalistic weekly the *Spectator* on March 16, "that Captain Scott may have reached the South Pole before Captain Amundsen, for although Captain Amundsen found no traces of his having done so it is to be remembered that the Pole is a vague point, and it would be possible in a region of mists and snowstorms for such fragile tokens as it is possible to leave behind them to be totally obscured."[32] Of course, Amundsen reported clear skies and fair weather at the pole.

Major British newspapers took a similar tack. "Even Sir Ernest Shackleton . . . declares that if Capt. Scott has reached the pole, it is probable that he would not hurry home," the *Evening Standard* reassured readers on March 9. "It is to be remembered that, unlike the Amundsen, the Scott expedition had much scientific work to perform." On the same day, an editorial in the *Times* of London commented, "We have still to hear the story of Captain Scott's expedition during the Antarctic summer; and it is by no means unlikely that he also succeeded in his chief endeavor, and, indeed, possible that he reached the Pole before December 14." Adding a gratuitous snub, the *Times* said of Amundsen's efforts, "So far as has appeared these results from a scientific point of view are disappointing." Even the Norwegian's voice in London, the *Daily Chronicle,* wrote, "England will wait most anxiously for news of the Scott expedition. Though robbed of its crowning glory, geography and science will undoubtedly profit from it."[33]

With no word from Scott, Amundsen set about cashing in on his triumph. While *Fram* sailed for Argentina, home of Pedro Christophersen, a wealthy Norwegian émigré who had helped provision the expedition, its leader lectured in Australia and New Zealand. Amundsen then followed *Fram* to South America, where he gave more lectures and completed his book about the expedition, *The South Pole,* for which he received a record advance in Norway. With his Arctic drift research on hold and *Fram* remaining in Buenos Aires, Amundsen embarked on a European lecture tour. "Mummy, is Amundsen a good man?" Kathleen Scott recorded her two-year-old son asking on March 11, 1912. "Amundsen and Daddy both got to the Pole. Daddy has stopped working now."[34] The boy was nearly right.

The British received a second blow on April 1, 1912, when further news dashed all hope that Scott had beaten Amundsen to the pole. *Terra Nova* had just returned to New Zealand after a second summer at the British base on Ross Island. The latest word from Scott, sent on January 3 to the main base with the final returning support team and conveyed by *Terra Nova* when it sailed north for the winter in early March, had the polar party 150 miles from the pole but otherwise in good shape. By this date, of course, the Norwegian party was partway back. "What, then, was our consternation when tidings were flashed over the world that Captain Roald Amundsen had been at the Pole?" the London literary magazine the *Bookman* asked. "Naturally enough, we resented in a way the wondrous good-fortune of one who . . . was a kind of interloper who had snatched the prize from the enclosing grasp of those who had more dearly won it."[35]

Early reports carried back on *Terra Nova* spoke of motorized tractors and ponies failing but Scott and four others persevering in the finest British tradition, man-hauling their sledges across the Polar Plateau. "It is unfortunate that we were unable to remain a week or ten days longer," a returning member of the *Terra Nova* expedition stated, "as we should almost certainly have back with us news that Capt. Scott had reached the Pole about Jan. 15. He and his companions probably returned to winter quarters before the end of March. No further news can be received from him until the *Terra Nova* again returns from the Far South about the end of next March."[36] Meanwhile, the expedition would endure a second winter in the Antarctic.

Science now became the only consolation for the British. "If less successful than Amundsen from the standpoint of adventure, Captain Scott and his colleagues promise much ampler scientific material," the left-leaning English weekly the *Nation* offered in April. "They have spent their energies in elaborate researches into climate conditions, geological studies, and inquiries into marine biology." The elite literary journal the *Athenaeum* added in Scott's defense, "His expedition had serious scientific objects, so the 'race for the South Pole,' imagined by some newspapers, never took place." Scott, the essayist suggested, was going to the pole in proper British fashion. Markham expressed the general view from the RGS: "There was no question of racing. The grand object was very far from that. It was valuable research in every branch of science. Capt. Amundsen's plan was different. He conceived of a dash for the south pole without Capt. Scott's knowledge." Similar comments appeared in other British journals. "We offer to Captain Scott and the other members of the British Antarctic expedition the thanks of the scientific world for the attention being given to systematic observations, which are of far greater value than the attainment of the south pole," the science journal *Nature* observed on April 4.[37]

Amundsen's decision to begin a European lecture tour while Scott remained in the field put the RGS in an awkward position. Viewing him as a base record-chaser who had deceived their man, many RGS leaders did not want to honor Amundsen but not doing so might look petty. After all, the society had feted Peary for reaching the North Pole even though he too had done little science and his claim was shrouded in controversy. At least no one doubted that Amundsen reached a pole. RGS President Lord Curzon, an explorer of central Asia who once served as British viceroy of India, announced the invitation in May at an RGS anniversary dinner attended by Prime Minister H. H. Asquith. While praising Scott's science and questioning Amundsen's honor, Curzon noted that these concerns "should not deter them from recognizing brave and adventurous achievement wherever accomplished." A more "hearty and enthusiastic welcome," he suggested, would await Scott.[38] To underscore the point, Amundsen's lecture was scheduled for Queen's Hall, a much smaller venue than the Royal Albert, where the RGS received Peary in 1910 and where it would surely receive Scott. Markham added to the insult at the 1912 meeting of the British Association for the Advancement of Science, where he gave a keynote address on Antarctic discovery that did not even mention Amundsen. He would only cover "true Antarctic expeditions," Markham said, not "mere dashes" to the pole. His lecture hailed Scott's work.[39]

A proud man sensitive to such slights, Amundsen canceled all plans to lecture in England. Fearing the impact on bilateral relations, King Haakon VII of Norway prevailed on Amundsen to go. "Personally I would have preferred to abstain, but when the king wishes it what else can I do?" Amundsen acknowledged privately. The encounter with the RGS was set for November 15. As the date approached, Markham counseled Kathleen Scott not to attend the lecture and wondered aloud how those RGS leaders who did attend could speak to her "after shaking hands with Amundsen." She returned her priority ticket for the event but slipped quietly into a top gallery. "Amundsen's speech was plucky and modest but dull, and of a dullness!" she wrote in the diary she still kept for her husband. "He did not mention you except just to say you were at McMurdo Sound." She noted with evident satisfaction that "there was scarcely anyone on the platform" with Amundsen except Curzon; Charles Darwin's explorer-son, past RGS president Leonard Darwin; and Scott's British rival, Shackleton. True to his word, Markham stayed away.[40]

As Kathleen observed, Amundsen gave a modest account that avoided contrasts with Scott. He offered due credit to his sled dogs. Curzon was not so discreet. In introducing Amundsen he spoke more warmly of Scott, "whose footsteps reached the same Pole, doubtless only a few weeks later than Amundsen, and who with unostentatious persistence, and in the true spirit of scientific devotion, is gathering in, during the absence of three years, a harvest of scientific spoils, which when he returns will be found to render his expedition the most notable of modern times." In his closing remarks following the banquet, Curzon added, "I almost wish that in our tribute of admiration we could include those wonderful good-tempered, fascinating dogs, the true friends of man, without whom Captain Amundsen would never have got to the Pole."[41] Then, as Amundsen remembered it, Curzon concluded with the phrase, "I therefore propose three cheers for the dogs," and turned toward the Norwegian.[42]

Amundsen never forgave Curzon's closing comment, which he

considered a "thinly veiled insult." He lectured throughout Britain with mixed success for another month. When the gate at some of the lectures in England could not cover his fee, Amundsen insisted on full payment even from charities that had organized the lectures to raise funds. "I won't let these damned English off one single penny," Amundsen wrote to his brother. "I will not yield a single point to this 'plum pudding nation.'"[43] Much larger audiences awaited him in continental Europe and the United States.

For their part, the British eagerly awaited a different explorer. "They must look forward to next spring," Markham told a huge crowd in Dundee, "when the whole country would welcome the return of Captain Scott, the greatest of all Polar explorers, and hear of the geographical achievements of himself and his gallant companions with the deepest interest and with well-founded national pride." A 1912 British children's book on polar exploration agreed. "Our own Captain Scott is rapidly approaching the goal," it assured readers, "and we sincerely trust that, though not first in the race—and, indeed, the obtaining of scientific results, not mere swiftness of march, was his aim—he too may win through."[44] Early in 1913, the prime minister assured Kathleen Scott that her husband would receive a peerage for his work in Antarctica. The honor only awaited his return. Then Scott, a gifted writer and speaker, would tell a story that would dwarf anything Amundsen could write or say. Drawing on more than a century of British experience in the Antarctic and a dozen years of focused research on Ross Island and Victoria Land, including his two grand expeditions there, he would spin them a tale of science, empire, and adventure at the world's end.

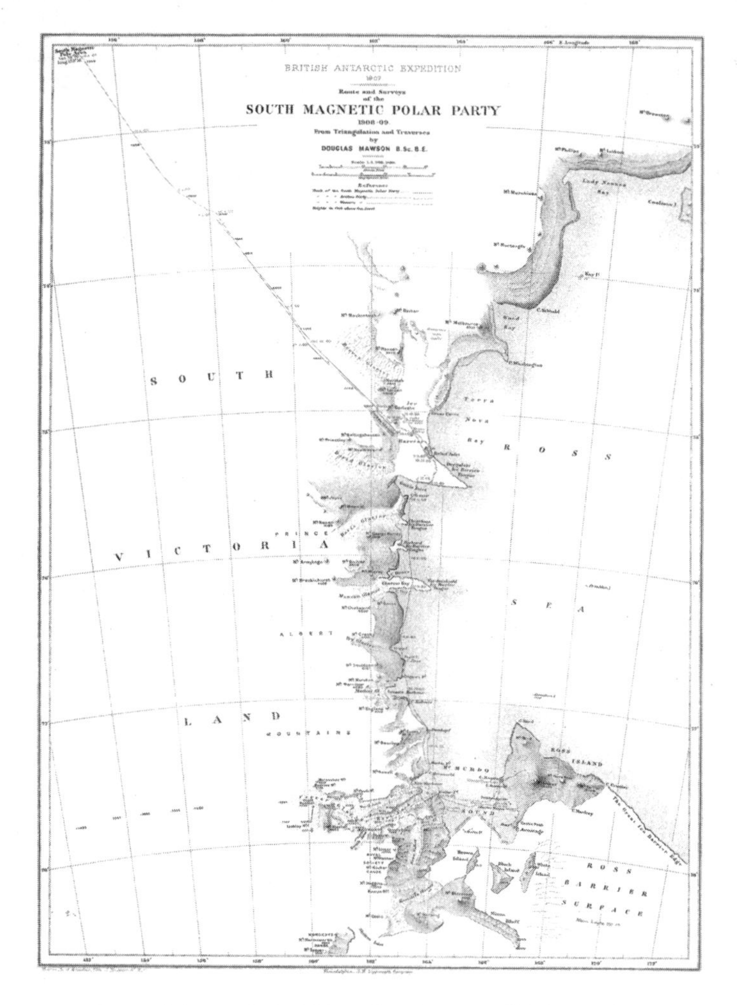

Route of Edgeworth David, Douglas Mawson, and Forbes Mackay to the South Magnetic Pole (1908–9), from Ernest Shackleton's *Heart of the Antarctic* (Philadelphia, 1909).

CHAPTER 2

A Compass Pointing South

"WHEREAS IT HAS BEEN REPRESENTED TO US that the science of magnetism may be essentially improved by an extensive series of observations made in high southern latitudes," the lords commissioners of the Admiralty, the Second Earl of Minto and the Third Baronet of Paglesham, wrote in 1839 to Captain James Clark Ross, "you will proceed direct to the southward, in order to determine the position of the magnetic pole, and even to attain it if possible, which it is hoped will be one of the remarkable and creditable results of the expedition."[1] The commissioners' orders went on to specify the parameters of Ross's four-year voyage of Antarctic research, which opened the way and set the standard for Scott. They made no mention of colonial acquisition and said little about geographical discovery. Further, despite beliefs that navigable seas might lie beyond the floating pack ice known to surround both polar regions, the orders spoke only of seeking the South Magnetic Pole, not the geographic one.

Improving the science of magnetism was the essential goal. "In the event of England being involved in hostilities with any other powers during your absence, you are clearly to understand that you are not to commit any hostile act whatsoever," Ross was ordered, "the expedition under your command being fitted out for the sole purpose of scientific discovery." If successful, the Admiralty lords assured him, the voyage "will engross the attention of the scientific men of all Europe." The British Antarctic Expedition of 1839–43 aboard HMS *Erebus* and *Terror* became one of the great voyages of discovery in British history—finding the Ross Sea, Victoria Land, and the Great Ice Barrier—but that was never its primary purpose. Terrestrial magnetism had become a fascination of the British and wider European scientific communities. This sent Ross south.

Perhaps coined as a sardonic allusion to the evangelical movements then bestirring England, the term Magnetic Crusade became associated during the 1830s with the orchestrated effort to chart the earth's magnetic field throughout the world. If it was a crusade, then the British Association for the Advancement of Science, or BAAS, was its church and Antarctica the Holy Land it sought to secure for science. In many ways, the crusade fit well with the scientific goals and imperial resources of early Victorian Britain.

Founded in 1831 as a semiprofessional, semipatrician organization of British cultivators of science, the BAAS embraced the utilitarian, fact-gathering ideal of Baconian research. By 1837, Charles Dickens was satirizing it as "the Mudfog Society for the Advancement of Everything" due to its unfocused and often uncritical attention to virtually any topic in science. To many of the BAAS's early leaders, terrestrial magnetism appeared to offer an ideal subject for their consideration: at once practical, theoretical, and in need of more observed facts from multiple sources. "The trophies of Newton, won at the end of the seventeenth century, made it impossible for the physical philosophers of the eighteenth not to attempt new victories in the application of mechanical principles to the phenomena of the material world," the Cambridge polymath William Whewell stated in his 1835 report to the association.[2] As Whewell saw it, Newton had used seemingly erratic observed planetary locations to produce a mathematical law of gravity that accounted not only for them but also for the movements of falling and orbiting bodies generally. Whewell, the Anglo-Irish geophysicist Edward Sabine, and other BAAS leaders viewed terrestrial magnetism as being on the cusp of a Newtonian moment.

European navigators had used magnetic compasses since the Middle Ages, and as early as 1600, the British natural philosopher William Gilbert had shown that a compass needle points north because the earth itself is a magnet. Compass needles do not point due north, however, but instead swing horizontally toward (though not always directly at) a North Magnetic Pole. Long-distance navigation by compass required knowing the precise variation (or declination) between the geographic north and the magnetic north. This declination varied by place and over time with no known laws explaining the changes. Further, as it approaches a magnetic pole, a magnetized needle dips (or inclines) vertically toward the pole.

If science could discover a regular law of declination and inclination, the great English astronomer Edmond Halley recognized in the late 1600s, navigators could use it to determine latitude and longitude at sea. Latitude was easily calculated from the sun's altitude at noon, but a compass method could help in cloudy weather. Determining longitude at sea proved an intractable puzzle until the English horologist John Harrison developed a reliable marine chronometer in the late 1700s, but even then an alternative compass method offered many advantages. The navigational value of knowing longitude and latitude appeared so great that, in 1699, the Admiralty gave Halley, a civilian, command of a navy vessel to measure magnetic variation and dip throughout the Atlantic Ocean to the high southern latitudes. It was the first purely scientific voyage by a British navy ship, and it ultimately reached past 50° south.

Charts of observed magnetic declination and inclination, with their gracefully curved isogonic and isoclinic lines, had limited lasting value because these factors change over time. As some sailors discovered through shipwreck, magnetic lines move—and straying off course by even a few minutes of arc could prove fatal. The earth's magnetic field also changes in intensity over time and place, which further confused its use in navigation. For over a generation after Halley's pioneering work, leading researchers virtually abandoned the field as hopelessly complex.

By the early 1800s, some physicists and naturalists had returned to the topic in the hope that, with sufficient data collected over an extended time from across the globe, they could discover regular laws of short-term (or diurnal) and long-term (or secular) change in terrestrial magnetic variation, dip, and intensity. In an era of expanding international trade, knowing these laws could improve the accuracy of navigation, particularly in the increasingly important sea lanes of the high southern latitudes—the so-called Roaring Forties and Furious Fifties—where magnetic readings were highly unreliable. The independent discovery during the early 1800s of relationships between magnetism, electricity, and mechanical motion gave reason to believe that any basic findings about the earth's magnetic field might also have practical value for communications and industry. Finding the fundamental laws of terrestrial magnetism became the object of the BAAS's Magnetic Crusade, which inspired Ross's expedition to Antarctica.

Despite its early prominence in the field, by 1830 Britain had fallen far behind its Continental rivals in magnetic research. Working mostly in Paris from 1804 to 1827, the charismatic German naturalist Alexander von Humboldt developed and popularized his theory that various dynamic cosmic and terrestrial forces continually affect the earth's magnetic field, causing it to change over time and place. Beginning in the 1820s, based on his theory that terrestrial magnetism emanated from stable features in or on the earth, the renowned German mathematician Carl Friedrich Gauss attempted to deduce regular periodic cycles of long- and short-term magnetic change from historical records. Both Humboldt and Gauss pleaded for more data, leading to the founding of magnetic observatories in France, Germany, and, under the direction of Adolf Kupffer, across the vast Russian Empire. Before 1830, some independent British researchers joined the cause, and taking magnetic readings became an ancillary activity on British Arctic expeditions. The highlight of this early effort came in 1831, when the dashing young navy commander James Clark Ross—widely regarded as the best-looking officer in the fleet—reached the North Magnetic Pole, where he triumphantly recorded a virtual 90° needle dip. This well-publicized episode helped stir popular support for magnetic research in Britain.

At its first annual meeting, held in 1831, the BAAS called for "a series of observations upon the *Intensity of Terrestrial Magnetism in various parts of England*." A year later, the association's second annual report added the lament, "When Baron Humboldt boasted to the French Academy of the wide distribution of his 'maisons magnetiques,' or magnetic observatories, from Paris, the center of civilization, to the wilds of Siberia, . . . it is a humiliating fact that he could not with truth have mentioned Britain as possessing a solitary establishment of this description." In its fifth year, the association voted to ask the government to erect magnetic observatories throughout the empire and to dispatch "an expedition into the Antarctic regions . . . with a view to determine precisely the place of the Southern Magnetic Pole or Poles, and the direction and inclination of the magnetic force in those regions." The idea of multiple poles derived from a hypothesis advanced by Halley and still favored by Sabine that two magnetic poles, plus a magnetically significant midpoint between them, existed in each hemisphere. They thought that the interplay of these points might explain the variability in the earth's magnetic field.[3]

Forewarned that the government was not yet ready to respond favorably, the association did not formally advance its plea for publicly funded magnetic research in 1835. Instead, to boost its case, it secured an endorsement for the proposal from Britain's senior science organization, the Royal Society, as well as a letter of support from Humboldt and the active involvement of the well-connected astronomer John Herschel.

In 1838, the association renewed its call for a network of government-funded magnetic stations in Britain and its colonies. The observatories would join with European facilities to make coordinated and, on certain "term days," simultaneous readings of magnetic "horizontal direction, dip, and intensity, or their theoretical equivalents" around the globe over a period of years. The new resolution again urged the navy to send an expedition to explore high southern latitudes, with a focus on the regions southeast of Australia where, according to Gauss, the South Magnetic Pole should lie. "The magnetism of the earth cannot be counted less than one of the most important branches of the physical history of the planet we inhabit," the gentleman geologist Roderick Impey Murchison noted in his 1838 presidential address to the association, "and we may feel quite assured, that the completion of our knowledge of its distribution on the surface of the earth would be regarded by our contemporaries and by posterity as a fitting enterprise of a maritime people." Ross, he said, should lead the Antarctic expedition.[4]

Herschel, already famous for his catalogue of stars in the Southern Hemisphere, stepped forward to offer the resolution on magnetic research at the association's 1838 annual meeting. Individual researchers could make magnetic observations in many places, he argued, but only a navy expedition could reach the Antarctic. "The seas, which must be the scene of inquiry," Herschel noted, "are visited by no commercial enterprise, and traversed by no casual vessels belonging to the British or any other navy. Nevertheless, it is vain to hope for a complete magnetic theory till this desideratum be supplied. The unsymmetrical form of the magnetic curves will baffle every attempt to reduce them under general laws, and will remain, as at present, an object of idle wonder, until these points, which may be looked upon as the keys to the enigma they offer, shall have been ascertained."[5]

Association annual meetings having become major public events in Britain, Herschel took the opportunity to address the sensitive issues of taxpayer cost and government priorities. Should anyone question spending public funds for "merely theoretical research," Herschel answered that "it is true [that the voyage] *is* to perfect a theory, but it is a theory, pregnant, as we see, with practical applications of the utmost importance." Declaring terrestrial magnetism the branch of science that "at the present moment, stands nearest to the verge of exact theory," he asserted that "great physical theories, with their trains of practical consequences, are preeminently national objects, whether for glory or for utility."[6]

Herschel also chaired the Royal Society committee charged with reviewing the proposal. "The changes in the position assumed by the [compass] needle at any particular point on the Earth's surface might be conceived as resulting from regular laws of periodicity," Herschel wrote in his committee's report. These periodicities could be determined by the coordinated collection of global data on lines of magnetic variation, dip, and intensity, "especially in the antarctic seas." Concluding "that a correct knowledge of the courses of these lines, especially where they approach their respective poles, is to be regarded as a first and, indeed, indispensable preliminary step to the construction of a rigorous and complete theory of terrestrial magnetism," the committee recommended, and the Royal Society resolved, that the British government should send a navy expedition to the Antarctic. The British scientific community stood united behind the Magnetic Crusade.[7]

Herschel turned his attention to lobbying the government. "Dined today with the Queen at Windsor Castle where had much conversation with Lord Melbourne [the prime minister], about the projected South Polar Expedition," he jotted in his diary for October 15, 1838.

Herschel also joined Sabine and Ross in lobbying the Admiralty. Science won the day. On September 30, 1839, *Erebus* and *Terror* departed from Britain with equipment for new magnetic observatories at the island of Saint Helena, the Cape of Good Hope, and Tasmania plus instructions to chart terrestrial magnetism in the "antarctic seas" from the Kerguelen Islands in the west to the Falklands in the east and south to the Magnetic Pole.[8]

Incoming BAAS President W. Vernon Harcourt, an aristocratic Anglican cleric with a passion for science, met with Ross shortly before the expedition set sail. "We sat down, Gentlemen, before his chart of the Southern Sea, and the unapproached pole of the earth," Harcourt reported at the association's 1839 annual meeting. "He put his finger upon the spot which theory assigns for the magnetic pole of verticity, corresponding to that which he had himself discovered in the opposite hemisphere. . . . I must confess, Gentlemen, it felt as one of the white and bright moments of life: such a conversation, at such a moment, with [such] a man." Harcourt called it "the most important and the best-appointed scientific expedition which ever sailed from the ports of England." Considering the expedition as part of a global Magnetic Crusade, Whewell expressed the feelings of many of its champions when he wrote, "The manner in which the business of magnetic observation has been taken up by the governments of our time makes this by far the greatest scientific undertaking which the world has ever seen."[9]

Ross's Antarctic expedition far exceeded any realistic expectations placed on it, and did so mostly in its first Antarctic summer. After establishing fixed magnetic observatories at Saint Helena, the South African cape, and Tasmania and taking its own regular magnetic readings in the Southern Hemisphere for over a year, the expedition broke through the ice pack south of New Zealand on a compass line for the South Magnetic Pole. No ship had ever penetrated this dense mass of broken sea ice. "After about an hour's hard thumping," as Ross described it, on January 5, 1841, the expedition found open sea beyond the pack and continued south. "We now shaped our course directly for the Magnetic Pole," which at this point meant sailing in a southwesterly direction. "Our hopes and expectations of attaining that interesting point were now raised to the highest pitch, too soon, however, to suffer as severe a disappointment."[10]

Six days later, the expedition encountered an unexpected obstacle in its path to the west: a vast expanse of land. "Although this circumstance was viewed at the time with considerable regret, as being likely to defeat one of the more important objects of the expedition, yet it restored to England the honour of the discovery of the southernmost known land," Ross noted in his report to the Admiralty. "It rose in lofty mountain peaks of from 9,000 to 12,000 feet in height, perfectly covered with eternal snow; the glaciers that descended from near the mountain summits projected many miles into the ocean."[11] Fittingly, Ross named many of the highest peaks for leaders of Britain's Magnetic Crusade: Sabine, Herschel, Whewell, Murchison, Harcourt, and more.

Blocked by the projecting ice from reaching the mainland, Ross claimed possession of it for Britain by briefly planting the nation's flag on an offshore islet. "The island on which we landed is composed wholly of igneous rocks, numerous specimens of which, with other imbedded minerals, were procured," Ross reported to the Admiralty. "Inconceivable myriads of penguins completely and densely covered the whole surface of the island," he added in his published account, "attacking us vigourously as we waded through their ranks, and pecking at us with their sharp beaks, disputing possession." In a flourish of unbridled imperial Victorian optimism, Ross named the place Possession Island and speculated that its deep beds of penguin guano "may at some period be valuable to the agriculturalist of our Australasian colonies." He called the newfound region Victoria Land.[12]

Thinking the sighted land might be merely a large island blocking his western advance, Ross attempted to reach the magnetic pole by sailing south along its coast. Regularly recording needle dips in excess of 88° at various points—which Ross estimated to put *Erebus* and *Terror* within 160 nautical miles of the magnetic pole—the expedition gradually worked its way south of its goal and far beyond the previous Farthest South attained by any vessel. It became apparent that, if Ross's estimates proved correct, the magnetic pole lay on land and either in or beyond the rugged range of coastal peaks. Much to his frustration, the ships were sailing past the magnetic pole without getting closer to it. By this time, the expedition was somewhere between the magnetic and geographic poles, and the ships' compasses showed it heading north rather than south.

"Still steering to the southward," Ross reported to the Admiralty, "a mountain of 12,000 feet above the level of the sea was seen emitting flame and smoke in splendid profusion." He named this volcano Mount Erebus for his flagship. Ross later commented, "The discovery of an active volcano in so high a southern latitude cannot but be esteemed a circumstance of high geological importance and interest, and contribute to throw some further light on the physical construction of the globe."[13] Erebus would become a prime objective for later researchers, including geologists on Shackleton's *Nimrod* expedition. Ross named a nearby dormant volcano Mount Terror for his other ship.

Not long after sighting Mount Erebus, Ross encountered a second obstacle: a massive wall of ice that blocked any further progress south and dashed his hope of finding a southern route around Victoria Land. "This extraordinary barrier presented a perpendicular face of at least 150 feet," Ross wrote, "completely concealing from our view every thing beyond it, except only the tops of a range of very lofty mountains in the S.S.E. direction."[14] After sailing east along the Great Ice Barrier for more than three hundred nautical miles, the explorers returned to the Victoria Land coast and, with winter bearing down, retreated to Tasmania, where the island's colonial governor, the former Arctic explorer John Franklin, greeted them warmly. He materially aided the expedition's efforts to construct the Rossbank magnetic and meteorological observatory in Hobart.

On the return voyage to Tasmania, Ross made a point of investigating the spot where Gauss's mathematical theory placed the Magnetic South Pole. "Having obtained all the observations that were necessary to prove the inaccuracy of that supposition," Ross reported, "we devoted some days to the investigation of the lines of no variation" before sailing on to Tasmania.[15] Throughout the expedition, *Erebus* and *Terror* served as floating observatories that systematically recorded magnetic variation, dip, and intensity in coordination with the international network of fixed stations. The sheer quantity of its magnetic readings was stunning, and their quality was better than anything previously taken at sea.

Although conducting magnetic observations was the expedition's principal objective and geographical discovery became its greatest legacy, the Royal Society also had sent detailed instructions for research in other fields. Little of the other work had immediate or lasting significance. Indeed, Ross only reprinted that part the instruction manual dealing with terrestrial magnetism in his account of the voyage. Except for Joseph Dalton Hooker, *Erebus*'s young assistant surgeon who signed on specifically to collect plants, and perhaps Ross, who had mastered magnetism and would became a student of marine invertebrates, none of the ships' officers and surgeons were trained scientists. Hooker published an important book on plants collected during the expedition that included pioneering observations on marine diatoms. The expedition launched his remarkable career in botany, but otherwise its scientific significance rested on its findings in magnetism and geography.

Erebus and *Terror* returned south for two more Antarctic summers before heading back to England. These years contributed little to the expedition's findings. During the second summer, a potentially disastrous collision between the two ships as they sought to elude an iceberg led to a prolonged layover for repairs in the Falklands, where research went on at a relaxed pace. "I never met with such devotees of science," one Falkland Islander noted. "You would be delighted to see Captain Ross's little hammock swinging close to his darling [geophysics] pendulum, and a large hole in the thin partition so that he may see it at any moment." Irony aside, systematic magnetic observations continued. On the ships' return, the *Times* of London gushed, "The acquisitions to natural history, geology, geography, but above all towards the elucidation of the great mystery of terrestrial magnetism, raise this voyage to a pre-eminent rank among the greatest achievements of British courage, intelligence, and enterprise."[16]

Despite their quantity and quality, the expedition's magnetic observations did not resolve the basic enigma. Sabine used them and supplementary ones amassed from the southern Indian Ocean by the 1845 voyage of HMS *Pagoda* and at observatories near Cape Town and Hobart to prepare the first complete map of magnetic lines for the Southern Hemisphere. But the collected data did not guide him to any predictive laws of magnetic change. "All attempts," he wrote, "that have hitherto been made to connect the secular magnetic change with any other physical phenomena, either terrestrial or cosmical, have signally failed."[17]

Five decades after Ross's voyage, researchers again began agitating for fresh magnetic readings from high southern latitudes, leading directly to Scott's *Discovery* expedition. Its scientists would join forces with those from other lands in a second outburst of Antarctica research. For a season, however, interest turned back to the North Polar Regions. In 1845, after Ross declined a commission to take *Erebus* and *Terror* to the Canadian Arctic for magnetic research and a possible transit of the Northwest Passage, the Admiralty recalled Franklin for the task. Both ships became icebound on his ill-fated journey, and no one returned. After leading an unsuccessful search effort, Ross retired to study the marine invertebrates collected on his grand Antarctic voyage.

The gracefully curved lines of the isogonic and isoclinic charts so laboriously produced by the Magnetic Crusade during the 1840s gradually became outdated. There was still no theory to predict changes in the earth's magnetic field. While the surviving network of fixed magnetic observatories supplied sufficient readings to adjust the charts for the middle latitudes, and an ongoing series of Arctic expeditions added enough data for researchers to track changes in the far north, magnetic lines in the far south again became a mystery. No observatories existed to monitor fluctuations in magnetic variation, dip, and intensity for the deep southern sea lanes, despite their steadily increasing use as Britain's colonies in Australia, New Zealand, and South Africa grew during the late 1800s. Further, with more ships being built of iron and steel, which distort magnetic readings, fewer of them could supply reliable magnetic observations.

Readings taken in the 1870s during the worldwide voyage of the British oceanographic research ship HMS *Challenger,* which briefly sailed south of the Antarctic Circle below the Indian Ocean, suggested that magnetic lines in the Antarctic region had changed dramatically since Ross's day. The South Magnetic Pole, which researchers had come to see as an area of land rather than a geographic point, appeared to have migrated northward. Navigators could not rely on old charts. In some areas around Cape Horn, readings suggested that the secular change in declination exceeded ten minutes of arc per year. *Challenger* naturalist John Murray came back from the voyage calling for renewed magnetic observations from Antarctica, ideally by land-based physicists over an extended period. Soon after, Admiralty Hydrographer Fredrick J. Evans reported to the Royal Geographical Society, "Which ever way we look at the subject of the earth's magnetism and its secular change, we find marvelous complexity and mystery." Commenting on Evans's report in its official journal that "there was therefore a wide field for future research in a primary geographical subject," the RGS soon emerged as the chief champion of a new British expedition to Antarctica.[18]

By the 1880s, Britain had a new geopolitical rival. The disparate German states humiliated France in the Franco-Prussian War in 1871 and then unified into a single empire to become Europe's preeminent Continental power. The Reich then began to build a deep-water navy to challenge British dominance on the high seas, and to assemble colonies in Africa and the South Seas. Germans had long played a leading role in magnetic research, and following the deaths of Gauss and Humboldt, the mantle fell on Georg von Neumayer. Just as Humboldt and Gauss had inspired the founding of an international network of magnetic observatories during the 1820s, in the 1880s Neumayer began advocating a German national expedition to make extended magnetic observations in the Antarctic. Some Britons, threatened by rising German nationalism, wanted to reassert their nation's claim to the Ross Sea region. "Certainly fifty years is a long time for us to have totally neglected so vast and so important a field," Clements Markham said in his 1893 inaugural address as RGS president.[19] Britannia, which still ruled the southern waves, sought better navigational charts to extend its reign.

Lobbying for a British expedition began in 1885, when retired Admiral Erasmus Ommanney, knighted for his Arctic service during the Franklin searches, lectured the BAAS on the need for new Antarctic research. Citing Neumayer's work, Ommanney argued that "another magnetic survey is most desirable in order to determine what secular change has been made in the elements of terrestrial magnetism after the interval of forty years." A naval expedition should winter in the Antarctic much as his ships wintered in the Arctic, he asserted, and investigate all fields of polar science. "No man has ever wintered in the antarctic zone," Ommanney noted. "The great desideratum now before us requires that an expedition should pass a winter there, in order to compare the conditions and phenomena with our arctic knowledge." Steamships made it practical, he said.[20] Invoking his *Challenger* credentials, Murray seconded Ommanney's call, which led the association to name a star-studded Antarctic Committee that included Ommanney, Murray, Hooker, Markham, T. H. Huxley, William Thomson (Lord Kelvin), and two old Arctic hands, Leopold McClintock and George Nares. "If this country spent as much on the exploration of the South Pole as some of our little wars cost," Murray declared, "they could thoroughly explore the Antarctic region."[21]

By this time, Neumayer had orchestrated the formation—without British participation—of the International Polar Commission, or IPC, with representatives from nine continental European nations. Its founding followed the beginning of coordinated international magnetic research in the Subantarctic by German, French, British, and American expeditions during both the 1874 Transit of Venus and, eight years later, the first International Polar Year. Under the IPC's auspices, during the early 1880s, Germany established a magnetic station on the island of South Georgia, France opened one in Tierra del Fuego, and navy vessels from several nations began systemically collecting magnetic readings in the high southern latitudes. Britain maintained observatories near Cape Town and Melbourne but watched its lead in Antarctic magnetic research diminish.

Despite three years of campaigning by the BAAS and the support of various Admiralty officials and Australian colonies, the tight-fisted ruling Conservative government refused to back a costly new Antarctic expedition. This was decisive because, based on their personal experiences on prior polar voyages, promoters of a renewed effort believed that only the British navy could successfully mount a sustained scientific expedition in Antarctica.

Of all these early promoters, Markham proved the most dogged. He had served as a midshipman in the Franklin searches, where he enjoyed the camaraderie of man-hauled sledging, but subsequently lost interest in navy service though not in naval officers or polar exploration, which he supported throughout his life. Leaving the navy as a young lieutenant in 1851, he worked in colonial administration and participated in public and private expeditions to such diverse places as Peru, India, Ethiopia, and Greenland. By the time he joined the BAAS's Antarctica Committee in 1885, Markham had served as RGS secretary for over two decades and had established himself as a force within the elite circle that dominated British geographical exploration during the late Victorian era. He was determined that Britain would lead the way in Antarctic discovery, and as Scott later observed, "Sir Clements was favorably placed for carrying out his determination."[22]

Although his public pleas for Antarctic exploration followed the party line in stressing the value of South Polar magnetic readings for science, Markham had other interests in mind. Privately, he described the expedition's main goal as "the encouragement of maritime enterprise" by young naval officers and observed that "the same object would lead to geographical exploration and discovery." He went on to dismiss the expedition's stated scientific goals as window-dressing designed to gain support or, as he put it, "springs to catch woodcocks."[23]

Even after the BAAS's initial push for a national expedition faltered, Markham kept an eye out for naval officers to lead it. Among his many activities, he served on a board for training marine officers and often entertained cadets in London. By his own account Markham first spotted Robert Falcon Scott in 1887, when the young midshipman won a hard-fought cutter race in the British West Indies. "He was then 18, and I was much struck by his intelligence, information, and the charm of his manner," Markham later wrote. At a similar training race five years later, Markham picked out cadet Charles Royds, whom he described as resembling "my *beau ideal* of a good polar officer," and in 1894 he met Albert Armitage, a former cadet then serving as navigator on an Arctic expedition. When the *Discovery* finally sailed, these men were its three principal officers. "He liked boys," British writer Francis Spufford observed about Markham, "but if he acted on the homosexuality he kept buried beneath a respectable marriage and an array of academic honors, he did so far away from home. Certainly far away from the midshipmen of good family, the bright-eyed merchant marine cadets, whom he began to make his companions in his middle age."[24]

Elevation to the RGS presidency in 1893 gave Markham a position to advance his Antarctic agenda. "Articles in magazines had to be published, lectures to be delivered, circulars to be sent out," he later recalled of his labors in the post on behalf of a south polar expedition.[25] Among his first acts as president, Markham invited Murray to address the RGS on the need for Antarctic research, appointed an RGS Antarctic Committee and solicited support for the expedition from the Royal Society, which promptly endorsed its scientific aims but never warmed to Markham's idea of making the expedition mainly a naval exercise with geographical ends.

Despite his personal preoccupation with geographical discovery, the need for magnetic research featured prominently in Markham's public pleas for an Antarctic expedition. "The science of terrestrial magnetism is at a standstill for want of recent observations from the far south," he warned in one of several articles on the topic, "and it is from this point of view that an Antarctic expedition can be shown to be necessary." Navigators urgently needed this information. "With regard to the probable results of an expedition, the most important would be the benefit to the science of terrestrial magnetism."[26] Commenting on grand aims of coordinated magnetic research, Markham explained that "the object is to achieve a series of synoptic charts which will allow of the variations in the magnetic condition of the whole earth being traced in detail during a definite period, and so to provide the necessary basis from which alone the fundamental problems of terrestrial magnetism can be more clearly approached."[27]

As orchestrated by Markham, Murray delivered his 1894 RGS address on the need for Antarctic research to an upper-crust audience of British scientists, explorers, and navy officers, who cheered his remarks. In his speech, Murray set the parameters for what became the *Discovery* expedition: a three-year, land-based, government-sponsored mission to an unknown continent. Although it might be possible to sledge quickly across the Great Ice Barrier to the pole, he noted, "A dash to the South Pole is not, however, what I now advocate, nor do I believe that it is what British science, at the present time, desires. It demands rather a steady, continuous, laborious, and systematic exploration of the whole southern region with all the appliances of the modern investigator."[28]

Although Murray envisioned a complete scientific survey of the region, he also paid homage to the importance of magnetic research. To justify the expedition, for example, he quoted Neumayer: "Without an examination and a survey of the magnetic properties of the Antarctic regions, it is utterly hopeless to strive, with prospects of success, at the advancement of the theory of the Earth's magnetism." He added Ross's lament that, if the 1839–43 expedition had found safe harbor in a Victoria Land bay, the coveted magnetic pole "might easily have been reached by travelling-parties in the following spring."[29] As the enthusiastic response to Murray's address underscored, the magnetic pole and its properties, not the geographic one, remained the aim of British Antarctic exploration. "It was a great meeting," Markham crowed, "and Sir John Murray's address was eloquent and convincing."[30]

Markham chaired the London meetings of the 1895 International Geographical Congress as president of the host society. With Neumayer in the hall and terrestrial magnetism front and center, Markham used the occasion to focus international attention on the need for an Antarctic expedition. "It is our duty," he told the delegates in his opening address, "to show how absolutely necessary some portions of the work of such an expedition have become—for example, the execution of a magnetic survey in the deep south—and to arouse public opinion in favour of an expedition." Neumayer then made his own case to the assembled delegates. "With fervid words and earnest manner he pleaded the cause of Antarctic research, gaining the sympathy and applause of his entire audience," one popular magazine reported. Hooker and Murray followed Neumayer to the dais and warmly endorsed his remarks. By a unanimous vote, the congress resolved that "a scientific exploration" of Antarctica was "the greatest piece of geographical exploration still to be undertaken."[31]

The congress also heard from Carsten Borchgrevink, an ambitious, scientifically interested Norwegian who had sailed before the mast on a commercial whaling ship to the Antarctic. Earlier in 1895, he erroneously laid claim to being the first person to set foot on the Antarctic mainland when his ship put a small party ashore for less than an hour on Cape Adare at the Ross Sea's western mouth. "His appearance before the congress was in the nature of surprise, and a hum of appreciative expectation filled the great Institute hall when his presence was announced," a journalist commented. Knowing of the delegates' interest in an Antarctic land expedition, Borchgrevink assured them, "I made a thorough investigation of the landing-place because I believe it to be a place where a future scientific expedition might safely stop even during the winter months." Adding the mistaken claim that from Cape Adare "a gentle slope leads on to the great plateau of the South Victoria continent," he offered to return with a proper scientific expedition and take a sledge party up the plateau to the South Magnetic Pole.[32]

Borchgrevink's offer attracted the attention of British publishing magnate George Newnes, who thought that an Antarctic expedition might make good copy for his mass-market magazines. In return for first publication rights, Newnes underwrote the cost for a modest expedition that left London in 1899 aboard a converted whaler renamed *Southern Cross.*

Southern Cross deposited Borchgrevink, nine other men, their supplies, sledges, dogs, scientific instruments, and two small prefabricated huts at Cape Adare in February 1899. It retrieved the nine survivors eleven months later. As his magnetic and meteorological observers, Borchgrevink chose British navigator William Colbeck, who had taken a quick course in terrestrial magnetism at Britain's Kew Observatory, and the young Australian physicist Louis Bernacchi from the Melbourne Observatory, who served a similar role in the *Discovery* expedition. Despite its exposed location on an isolated peninsula and the death of its able zoologist, Nikolai Hanson, the expedition was the first to winter on land in Antarctica, and it returned with some valuable natural-history specimens and scientific observations. The mountains and water surrounding Cape Adare, however, frustrated every attempt to push into the interior toward the magnetic pole. Further, the expedition lost the distinction of being the first to winter below the Antarctic Circle because a Belgian research vessel carrying Roald Amundsen as first mate had become trapped in the late-summer ice on the other side of Antarctica a year earlier and remained there until the following summer. Both expeditions kept regular magnetic and meteorological records—the first made over an Antarctic winter.

Imperial British science could be fickle. Borchgrevink received more acclaim in 1895 for his unsubstantiated boast of being the first person to step on Antarctica, when that chance occurrence served to show that landing a party there was possible, than he did four years later for actually being the first to winter on the Antarctic mainland, when doing so prevented Scott and his men from claiming that honor. Markham, who had once hailed Borchgrevink as an Antarctic pioneer, later snubbed him in public and cursed him in private for diverting funds and attention from the *Discovery* expedition. Geography was a cut-throat enterprise in late Victorian Britain. Following his return, Borchgrevink was criticized for allegedly mishandling Hanson's notes and specimens and for not pushing with sufficient vigor into the interior toward the magnetic pole. Both of these supposed failings of his groundbreaking endeavor conveniently left more for Scott's all-British team to achieve.

By the time Borchgrevink returned to England in 1900, the British government had finally succumbed to prodigious lobbying by Markham and much of the nation's scientific establishment and had appropriated public funds for an Antarctic expedition. Even then, the government agreed only to match privately raised funds, most of which came from the British industrialist Llewellyn W. Longstaff, who responded to Markham's nationalistic pleas for British science with a large enough donation to ensure that the project would proceed. "I only hope that my action may result in important gains to science," Longstaff wrote to Markham, quoting back Markham's own motto about the effort. "I felt that 'It should be done and that England should (help to) do it.'"[33] Crucially for Markham's vision of the venture, the Admiralty also agreed to release a limited number of junior officers and crewmen for the expedition, which assumed a naval bearing even though *Discovery* sailed as a private yacht. But Ernest Shackleton, the expedition's charismatic third lieutenant, came from the merchant marine on Longstaff's recommendation. Chafing under naval discipline, he would become the only officer to threaten Scott's authority.

The expedition was costly because its organizers opted to commission a new ship designed for scientific research rather than use a converted Arctic whaler. Adding to the cost, an oversight committee composed of RGS and Royal Society members insisted on a wooden hull in age of iron ships to permit the taking of accurate magnetic readings at sea. "As the *Discovery* will have to do important magnetic work, a special magnetic observatory will be constructed and fitted on the upper deck," the *Times* of London reported. "Special care will be taken that no ironwork shall be used for any purpose within a distance of 30 feet of this observatory. Where metal must be used, it will consist of rolled naval brass."[34] The ship's auxiliary steam engine was placed as far as possible from the observatory. *Discovery* was a purpose-built ship and its purpose was magnetic research on open water, in pack ice, and when frozen in place. It cost over twice as much as a whaler, rolled excessively in heavy seas, and leaked badly.

The final nudge for the British government to support Antarctic research came in 1899, when the German Reichstag yielded to Neumayer's pleas by voting to fund an Antarctic expedition commanded by a scientist-explorer, Erich von Drygalski. It, too, sailed in a purpose-built wooden ship designed for magnetic research, the aptly named *Gauss*. Ultimately, Britain would not abandon this field to its rising rival. "The Nation's credit is at stake," Markham warned even as he urged that the two empires collaborate on a comprehensive Antarctic magnetic survey. "Shall we let strangers do England's work?" At the time, Britain and Germany were still probing their new relationship as potentially coequal European powers. "It must rejoice the hearts of all geographers," Markham proclaimed, "that the countrymen of Humboldt . . . should combine with the countrymen . . . of Sabine to achieve a grand scientific work which will redound to the honor of both nations."[35] At the 1899 International Geographical Congress in Berlin, delegates agreed that Britain would focus its efforts on two quadrants of the Antarctic south of Australia and the Pacific Ocean, which included the Ross Sea, while Germany would explore the little-known quadrant under the Indian Ocean. The final quadrant went to Sweden, which sent an expedition that contributed a coordinated set of magnetic observations despite having its ship crushed in the sea ice off the Antarctic Peninsula.

After establishing a base magnetic station far south in the Indian

Ocean on the deserted Kerguelen Islands in 1901, the *Gauss* expedition sailed toward Antarctica. But the ship became locked in sea ice more than fifty miles from the coast, and its scientists spent the winter taking magnetic readings and conducting such other research as their location allowed before the ship broke free and returned north in 1903. Drygalski was forced to abandon his earlier plan of attempting to reach the magnetic pole from the west and made only sporadic trips over the sea ice to the mainland.

During the same period, privately funded expeditions from Scotland and France explored the Antarctic Peninsula and Weddell Sea. Although they also took magnetic readings, these expeditions were not invited to participate in the coordinated magnetic survey built around the voyages of the British, German, and Swedish ships. When the Scottish expedition's proven leader, William Speirs Bruce, offered his ship's service for the survey, Markham—already furious with him for diverting private Scottish funds from the official British expedition—refused on the doubtful grounds that Bruce's vessel was unsuited for magnetic work. "If not specially built for magnetic observations," he wrote to Bruce, "your ship cannot co-operate with the British or German Expeditions in that way, which is the main point; so that there can be no cooperation on your part. The National Expedition is in serious want of more funds, and your proceedings are exceedingly harmful to it, by diverting all chance of help, so far as you can succeed in doing so, from Scotland."[36] The Scots placed their observatory on the South Orkney Islands, where it still operates.

A deeply divided joint committee of the RGS and Royal Society sent official instructions to the *Discovery* expedition that gave equal primacy to magnetic research and geographical discovery. "The objects of the expedition are (*a*) to determine, as far as possible, the nature, condition, and extent of that portion of the south polar lands which is included in the scope of your expedition; and (*b*) to make a magnetic survey in the southern regions to the south of the 40th parallel." With the RGS caring most about geography and the Royal Society primarily interested in magnetism, the instructions stressed that "neither of these objects is to be sacrificed to the other." As Markham later noted, "All mention of the [geographic] south pole as an objective was carefully avoided." Regarding the magnetic survey, the committee counseled Scott, "The greatest importance is attached to the series of magnetic observations to be taken under your superintendence, and we desire that you will spare no pains to insure their accuracy and continuity." In particular, the expedition was to take regular magnetic readings both as it sailed in the South Pacific or Ross Sea and on sledge journeys from an Antarctic base station after it landed. Armitage supervised magnetic research at sea; Bernacchi took charge on land.[37]

The Antarctic magnetic station consisted of two small asbestos-walled huts erected at Ross Island behind the expedition's main building upon a rocky peninsula that became known as Hut Point. By the shore of McMurdo Sound across from Victoria Land, this site served as the explorers' winter quarters. Official instructions called for a party to winter in Antarctica, with Scott free either to keep *Discovery* "in the ice" or to sail it north for the winter.[38] He opted to winter his ship at Hut Point. It remained frozen in place for twenty-four months before the sea ice broke out. The expedition's extended stay at Hut Point permitted the systematic collection of magnetic readings over a two-year period.

During their first year at Hut Point, in addition to making regular observations of variation, dip, and intensity, *Discovery*'s researchers coordinated with those on *Gauss* and at other deep-southern stations in taking simultaneous observations of declination, vertical force, and horizontal force every hour on twice-monthly term days and every twenty seconds for one hour on each of those days. Bernacchi, whom Scott privately characterized as "full of grand conceits and foibles but full also of puck and a hard worker," coordinated and largely conducted the effort. "For the ordinary routine work," Bernacchi noted, "there were self-recording instruments which required to be visited once or twice a day, but on the 1st and 15th of every month, the international term days, I had to visit the magnetic huts every two hours of the twenty-four, to change the photographic recording sheets." Because the unheated huts became buried in snow, which helped to keep conditions constant for the instruments, the inside temperature hovered around −30°F. Bernacchi's work resulted in a three-hundred-page book of magnetic readings published by the Royal Society in 1909 and used by physicists for a generation. "In general," Scott wrote to a Royal Society official on *Discovery*'s return to Britain, "I am ready to plead ignorance of the value of our Scientific work but when it comes to magnetic results I make bold to say that our light ought not to be dimmed by anything that other expeditions can produce."[39]

Locating the South Magnetic Pole remained a stated objective of the Antarctic survey. Unaware that Drygalski had abandoned his quest for it, the British expedition made its assault on the magnetic pole during its first full Antarctic summer. Already drawn to the idea of reaching the geographic pole, during the first summer Scott headed south across the Ice Barrier, with Shackleton and *Discovery* physician, artist, and budding zoologist Edward A. Wilson. But knowing the importance Markham placed on claiming the magnetic pole for Britain, on November 29, 1902, Scott dispatched a large team west under Armitage with the unrealistic goal of reaching it. At least, he reasoned, this Western Sledge Party might discover whether an ice sheet or coastline lay beyond the Victoria Land mountains. This in turn might resolve the geographical question, was Antarctica a continent or an archipelago of islands?

Because it remained icebound at Hut Point, far to the south and west of the magnetic pole, *Discovery* could not ferry Armitage and his men to a point on the same parallel with the pole. Yet the sea ice farther north looked too unstable for sledging. So the party crossed the stable ice near Hut Point and from there ascended a glacier between the mountains in the hope of finding a route north on the far side. "In all twenty-one souls went forth to try to surmount that grim-looking barrier to the west," Scott observed. In deference to Markham's armchair conception of this venture, which drew heavily on his nostalgic memories of the Franklin searches, they pulled their sledges without the aid of dogs. "The distance of Ross's magnetic pole from McMurdo bay and back could very easily be covered in three months," Markham had told fellow geographers in 1899, "without the cruelty of killing a team of dogs by overwork and starvation."[40]

Nothing in the sixty-year history of man-hauling sledges in the Arctic could have prepared Armitage and his men for the ordeal of pulling ten heavy sledges, loaded with two tons of supplies, equipment, and scientific instruments, across the mountains of Victoria Land. Armitage initially hoped to run the sledges on a glacier that cut through the range but, on inspection, that route looked impassably steep and icy. "It would be madness to attempt it," one officer advised him. Instead the party tried to pick a route between the peaks by relaying loads and using improvised block-and-tackle techniques to get the sledges up steep pitches and over outcroppings. Each false summit simply led to another until the party reached a dead end after gaining some six thousand feet in altitude. "To our intense disappointment, there was no route for sledges," Armitage wrote. "All our toil had been for nothing."[41]

Rather than give up, they went back down to try the glacial route, which proved almost as difficult as it looked. Every man showed signs of scurvy, and some were left behind. One suffered a near-fatal heart attack. Armitage once fell through a snow bridge and hung by his harness until rescued. Blizzards kept the party confined for days; icefalls, crevasses, and moraines blocked the way. "The worst feature of a pioneer journey inland is that, not knowing where the road leads to, it is impossible to proceed in thick weather," Armitage noted. After thirty-seven days, he finally summited with five men at about nine thousand feet and looked out over the Polar Plateau. "How far the ice-cap extended it was impossible to say," Armitage added, "we had to rest content with being the first men to have actually found it." Utterly beat, they turned back without further thought of the magnetic pole. Collecting their colleagues along the way, they reached Hut Point without losing anyone. "It was a piece of excellent pioneer work," Markham later declared. "The journey," Scott commented in his first-year's report to the Royal Society, "reflected great credit on Mr. Armitage." William Bruce hailed the effort as "almost superhuman."[42]

The Western Sledge Party showed a way through the coastal range and suggested what lay beyond. Its magnetic readings helped researchers compute the magnetic pole's location, which had migrated significantly since Ross's day. Further, with the expedition extended for a second summer, Scott used Armitage's route to reach the plateau but then, instead of turning north toward the magnetic pole, pushed west in a vain effort to cross Victoria Land. This left a second grand objective for Shackleton when he returned to Ross Island aboard *Nimrod* in 1908 with his own privately funded expedition.

Shackleton's *Nimrod* expedition was all about Firsts and Farthests. "I never saw anyone enjoy success with such gusto as Shackleton," RGS Librarian Hugh Robert Mill observed. "His whole life was to him a romantic poem, hardening with stoical endurance in adversity, rising to rhapsody when he found his place in the sun."[43] Although he had his eyes fixed on being first to the geographic pole, or at least getting farther south than he had with Scott in 1902, Shackleton recognized the long-standing scientific interest in reaching the magnetic pole and made that "first" a part of his expedition's mission. For the so-called Northern Sledge Party, whose journey to the magnetic pole would coincide with his own run south, Shackleton chose the expedition's fifty-year-old geologist, noted University of Sydney professor

T. W. Edgeworth David; David's former student Douglas Mawson, who taught mineralogy at the University of Adelaide but joined the expedition as its physicist; and the expedition's Scottish physician A. Forbes Mackay.

Placing David in overall command, Shackleton directed the party "to take magnetic observations at every suitable point with a view of determining the dip and the position of the Magnetic Pole. If time permits, and your equipment and supplies are sufficient, you will try and reach the Magnetic Pole." Reaching the magnetic pole was a highly ambitious goal for a three-person, two-sledge team pulling up to twelve hundred pounds over an unknown route in subzero temperature. The party would leave the expedition's winter quarters at Cape Royds, nearly twenty miles north of Scott's base at Hut Point, early enough in October to sledge over the sea ice to Victoria Land and then proceed north along the frozen coast to Terra Nova Bay. Including relay work, the outbound Ross Sea portion of the trip totaled 260 miles. Then the party would need to climb up to and cross a portion of the seven-thousand-foot-high Polar Plateau—a 520-mile-long round trip. They roughly knew these distances in advance from data gathered by the *Discovery* expedition. Shackleton also directed the party "to work at the geology of the Western Mountains" and "to prospect for minerals of economic value" in the snow-free "Dry Valley" discovered there by Scott in 1903. Curiously, it was the young physicist Mawson who favored geologizing in the nearby Dry Valley, while the older geologist David seemed hell-bent on reaching the magnetic pole.[44]

The Northern Sledge Journey was a forced march of foreseeable brutality. Early on, David suggested traveling on half-rations to lighten the load. "I cannot see how it is possible for us to reach the Magnetic Pole in one season under such conditions," Mawson complained in his journal. Instead, he reported telling the others near the

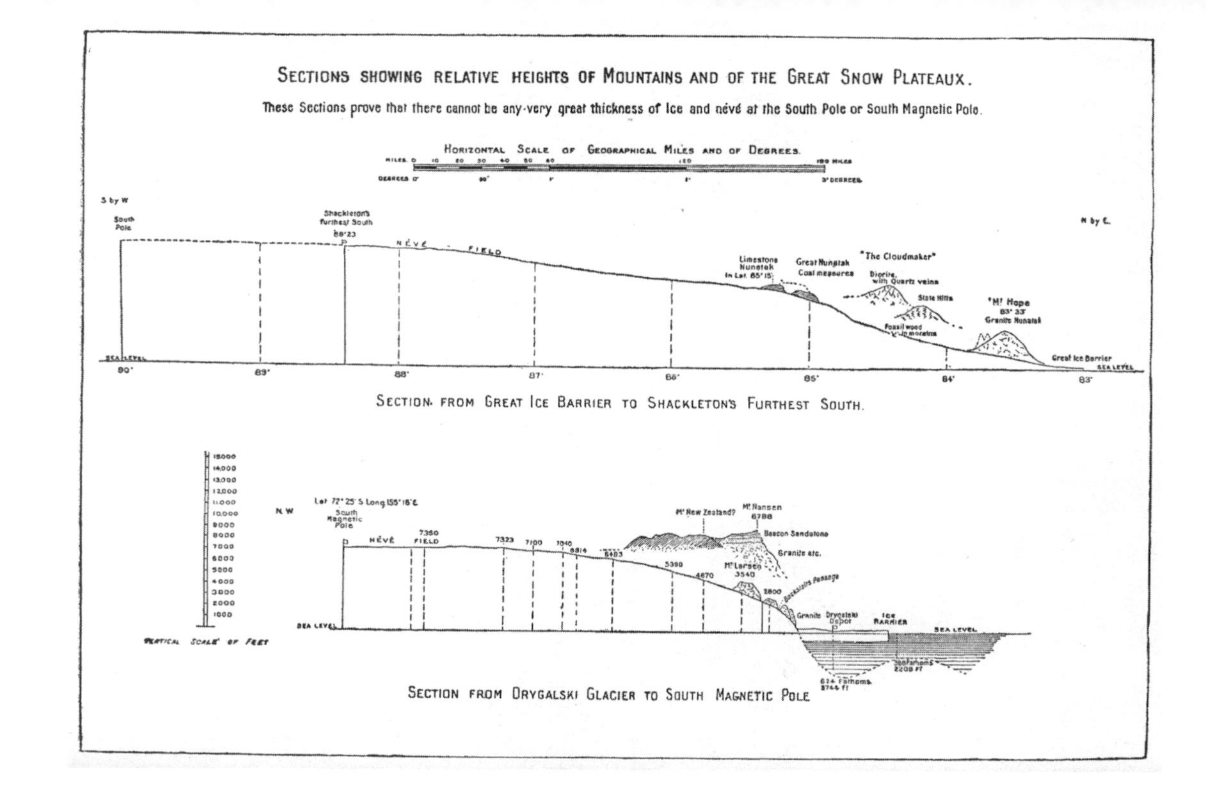

Sketch showing the elevation of mountains and the polar plateau crossed by the *Nimrod* expedition's Northern and Southern Sledge Journeys (1908–9), from Ernest Shackleton's *Heart of the Antarctic* (Philadelphia, 1909).

outset, "(1) We must give up all else this summer, (2) preserve about a full ration of sledge food for 480 m journey inland, (3) in order to make this possible we must live on seal flesh and local food . . . as much as possible." The trip would take over three months, he calculated. With the sea ice gone by the time they returned to the coast, they would have to hope that *Nimrod* came to retrieve them. Mawson wanted to give up the pole and concentrate on geology, but David refused. "I cannot do anything but agree and I give my whole power toward it," Mawson asserted in a private journal entry.[45]

Hunger stalked them. "About 18 days out I had a food dream," Mawson wrote. "Dreamt that we came upon a depot containing all sorts of choice delicacies." Instead, their overland sledging rations consisted mostly of pemmican and biscuits from which they made hoosh stew. While on the sea ice, they lived mainly on seals and penguins. "Young bull seal is always good—steak from loins, liver and blubber," Mawson noted. "The tongue we found good but kidney rubbery and useless. Cow seal in breeding season is to be avoided. . . . Have found the penguin liver the best of all yet." Still, the food was never enough and sometimes made them sick. "When feeding on seal meat," Mawson reported, "we opened out every hour at least." They planned future feasts while sledging. "We dote on what sprees we shall have on return—mostly run to sweet foods and farinaceous compounds. We don't intend to let a meal pass in after life without more fully appreciating it."[46]

After two months on the sea ice, the party began ascending a glacial incline toward the plateau. Blizzards became common, and the men regularly fell into crevasses. They set themselves a goal of covering at least ten miles per day, which could mean more than twelve hours of heavy pulling. They got little sleep. "A three-man sleeping-bag, where you are wedged in more or less tightly against your mates, where all snore and shin one another and each feels on waking that he is more sinned against than shinning, is not conducive to real rest," David commented in his journal. "The Prof is dreadfully slow," Mawson complained about the middle-aged David on New Year's Eve. "Food very little now as original supply has undergone two reductions." Worst of all, Mawson added, "Dip reading very little less than previous and very discouraging."[47] Frostbite, snow blindness, and bleeding lips plagued them; David stumbled often. On January 12, when they had reached the place where they thought Bernacchi's data placed the pole, they found that it had migrated northwest and they needed to go forty more miles. "This was extremely disquieting news, for all of us as we had come almost to the limit of our provisions," David wrote. "We decided to go on for another four days."[48]

On the fourth day, January 16, 1909, they reached the magnetic pole—or at least the vicinity of where Mawson estimated that the pole's mean position should lie. "The determination of the exact centre of the magnetic polar area could not be made on the spot, as it would involve a large number of readings taken at positions surrounding the Pole," he explained. As it was, he barely had time to rig the camera for a group photo. "Meanwhile, Mackay and I fixed up the flagpole," David wrote. "We then bared our heads and hoisted the Union Jack at 3:30 p.m. with the words uttered by myself, in conformity with Lieutenant Shackleton's instructions, 'I hereby take possession of this area now containing the Magnetic Pole for the British Empire.' At the same time I fired the trigger of the camera." They gave three cheers for the king and left. "It was an intense satisfaction," David observed, "to fulfill the wish of Sir James Clarke [sic] Ross that the South Magnetic Pole should be actually reached, as he had already in 1831 reached the North Magnetic Pole."[49]

Seventy years after Ross first set sail for it, British polar explorers had finally claimed this Antarctic goal, but scientifically it did not mean much. Magnetic poles hover in an area rather than reside at a point, and members of the exhausted party made no dip readings at the claimed site. They had left their magnetic instruments at the final outbound camp and, on the last day, simply sprinted toward their best guess as to the pole's position. Their compass's horizontal element had ceased working several days earlier, and the recorded dip never quite reached 90°.

David, Mawson, and Mackay then faced an agonizing 260-mile dash for the coast in an increasingly desperate attempt to catch *Nimrod* before it sailed north. They made it to the coast in less than three weeks despite frequent falls due to the steep ice and soft late-summer snow. Apparently brutalized by the effort, Mackay began kicking David to speed his pace and called him "a bloody fool" for falling. By the end, David was so disorientated that he relinquished command to Mawson. "He says himself that had he known the magnitude [of the task] he would not have undertaken it," Mawson wrote in his journal four days before the *Nimrod* rescued them.[50] Years later, both David and Mawson concluded that they had probably never quite reached the pole.

For the expedition, however, their journey was seen as a towering achievement. Bernacchi hailed it as "most valuable and important." When Shackleton fell short of reaching the geographic pole, the Northern Sledge Party's feat became nearly coequal with his own Farthest South. "I congratulate you and your comrades," King Edward VII telegraphed to Shackleton on the *Nimrod*'s return to New Zealand, "on your having succeeded in hoisting the Union Jack . . . within a hundred miles of the South Pole, and the Union Jack at the South Magnetic Pole." In the RGS's official journal, Mill depicted David's north as "no less interesting" than Shackleton's south.[51] In his widely popular book and lectures about the expedition, Shackleton, always a savvy promoter, gave almost as much attention to the Northern Sledge Journey as to his own southern one. Some muddled press accounts made it appeared that Shackleton had made both treks.

Britain's long Magnetic Crusade, begun in the first years of Victoria's long reign and wrapped in popular notions of science and empire, finally reached its Jerusalem with three cheers for the king somewhere near the South Magnetic Pole.

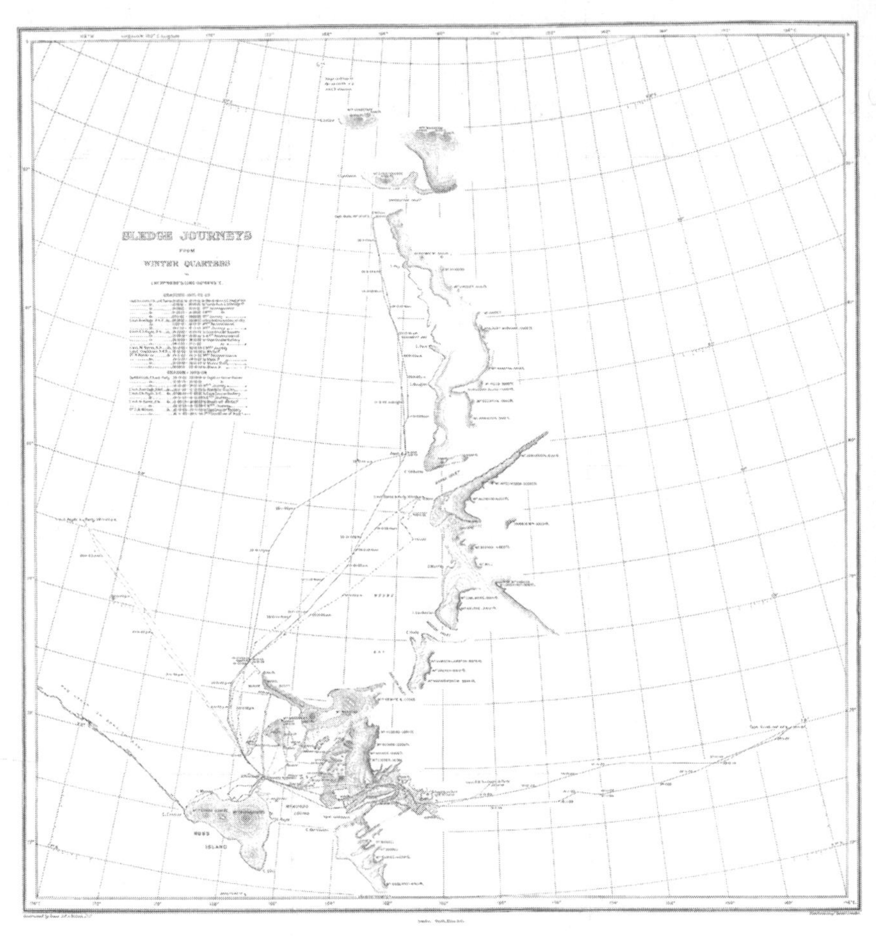

Discovery expedition sledging routes, south, west, and east (1901–4), from Robert Scott's *Voyage of the "Discovery"* (New York, 1905).

CHAPTER 3

The Empire's Mapmaker

SCIENCE CARRIED AUTHORITY IN THE ENGLISH-speaking world at the time *Discovery* sailed, and *Discovery*'s sponsors claimed the mantle of science for their work. Yet Victorians contested the meaning of science by ascribing it all manner of means and ends. In late-nineteenth-century Britain, virtually anything might be called scientific by someone interested in touting its virtues.

Although there was some common ground over what constituted good science in the context of Antarctic exploration, as drafted by its sponsors, the *Discovery* expedition's instructions gave equal emphasis to two forms of research from different scientific traditions. The Royal Society, which increasingly was composed of professional scientists committed to experimental research and the systemization of knowledge in physics, biology and the other natural sciences, called for an Antarctic magnetic survey. The instruction "to determine, as far as possible, the nature, condition, and extent . . . of the south polar lands," on the other hand, articulated the aims of the Royal Geographical Society, an association of amateurs devoted to continuing a tradition of mapping, fact gathering, and specimen collecting by explorers and naturalists in the service of empire. Clements Markham and other RGS leaders viewed their society's activities as scientific even though some in the Royal Society might dismiss them as craft or even hobby. At the RGS's urging, the expedition's instructions authorized "geographical discovery and scientific exploration by sea and land" in the Antarctic quadrants surrounding the Ross Sea, which at Markham's request were assigned to Britain by the 1899 International Geographical Congress.[1]

Neither the Royal Society nor the RGS wanted a record-chasing dash to the pole or mere adventure travel. Both demanded a thoroughly scientific enterprise, though they had diverging ideas of what that might mean. For the Royal Society, it typically meant the collection of data in the form of numerical tables or graphs for analysis by experts in Britain. In contrast, the meaning the RGS placed on such terms as "geographical discovery" and "scientific exploration" grew out of its seventy-year history during Britain's century of empire. A maker of maps, the society took an approach toward Antarctica that was largely shaped by its experience with Africa—an experience that, by 1900, had made geographical explorers into British national heroes.

Founded in 1830 as little more than a social club, lecture hall, and map room for English gentlemen interested in world travel, in the 1850s the RGS began to take a central role in shaping Britain's conceptions of geography when it welcomed David Livingstone back from his missionary travels and researches in southern Africa. The RGS had neither dispatched nor trained Livingstone for geographical work, but the society's able and ambitious president, the gentleman-geologist Roderick Impey Murchison, recognized the restless Scottish missionary's extraordinary gift for African fieldwork and helped mold him into the living exemplar of a scientific British explorer for a Christian imperial age. In an era when Britons wanted heroes, both men loomed large. The collaboration of Murchison and Livingstone in opening central Africa to Europe set the pattern that Markham and Scott would follow in Antarctica a generation later.

Before Livingstone, the RGS had served the interests of maritime travelers more than those of empire builders. Following the loss of the American colonies in 1783 and a long war with Napoleonic France ending in 1815, many Britons had little stomach for acquiring a global land-based empire. But Britain's navy ruled the seas, and its far-flung commercial interests included trading posts and whaling stations around the world. Beginning with James Cook's first voyage in 1768, the Admiralty launched expeditions to survey coasts, harbors, and seas for military and commercial purposes. Young naturalists destined for greatness, like Joseph Banks, Charles Darwin, Joseph Hooker, and T. H. Huxley, might accompany these voyages, but their fieldwork was limited by sailing schedules. Yet the international reach of British commerce facilitated the spread of British travelers, especially following the introduction of ocean-going steamships during the second quarter of the nineteenth century. Soon the world was teeming with Britons chasing remote destinations.

For most of its first two decades, the RGS remained an elite dining club where wealthy English adventurers, diplomats, and military officers swapped travel stories over exotic dishes. During the mid-1830s, under the leadership of Admiralty Second Secretary John Barrow—who showed greater interest in dispatching navy expeditions to the Arctic than in nurturing land exploration—the society settled into its roles as an association of well-heeled geography enthusiasts and a repository of maps, information, and instruments for explorers, travelers, and the military. Membership and financial resources dwindled throughout the 1840s until Murchison took charge in 1851 and quickly made the RGS into a command post for land exploration and global empire. Public lectures by returning explorers became its main drawing card, with priority access for members.

A field geologist who did important work in tracing the rock formations of Europe, Murchison presided at various times over the Geological Society of London, Britain's Geological Survey, and the British Association for the Advancement of Science as well as the RGS.

Geography for him was closely allied with geology in the scientific study of the earth, its features, and its peoples. Geological surveys looked below the surface to map the distribution and composition of mountain ranges and rock strata, with a practical eye for coal seams and beds of mineral-bearing rock. Physical geography focused on the surface by mapping topography, describing features, and plotting places by their longitude, latitude, and altitude. Human geography added the mapping and description of peoples and cultures. Murchison believed that both geological and geographical information could guide British colonial expansion and, when engagingly presented by intrepid explorers, also entertain domestic audiences. "His sophisticated manipulation of publicity techniques," wrote Murchison's biographer, Robert A. Stafford, "transformed the ailing Society into a theatre of national suspense, a company of talented adventurers purveying high drama in exotic settings."[2]

Still largely unmapped by Europeans beyond its coasts and of grave concern to British Christians opposed to the slave trade, Africa took center stage in the RGS theater. As early as 1852, Murchison speculated that an elevated central basin, filled with lakes and rivers, lay beyond the continent's forbidding rim of malarial coasts, barren deserts, and rugged mountains. The rivers running from the interior to the coast, particularly the Nile, which would connect these highlands to the Mediterranean, excited his interest as possible routes for British commerce and colonization.

"Wherever the sources of the Nile may ultimately be fixed and defined, we are now pretty well assured they lie in lofty mountains at no great distance from the east coast," Murchison declared in his 1852 RGS presidential address. "The adventurous travelers who shall first lay down the true position of these equatorial snowy mountains . . . and who shall satisfy us that they not only throw off the waters of the White Nile to the north, but some to the east, and will further answer the query, whether they may not also shed off some streams to a great lacustrine and sandy interior of this continent, will be justly considered among the greatest benefactors of this age to geographical science!"[3] While the RGS also backed expeditions to Asia, Australia, and other far-off regions during Murchison's sixteen-year tenure as president, which included all but five years from 1851 to 1871, Africa received its most sustained attention, and lectures on African Nights typically drew the largest crowds. Membership doubled and then doubled again.

Before turning to Livingstone, Murchison tapped the intrepid but irascible Richard Francis Burton and his sometime companion John Hanning Speke to seek the Nile's source in an area of east-central Africa unknown to Europeans but rumored to contain great lakes, high mountains, and a substantial native population exploited by Arab slave traders. This region especially interested Murchison and other British Christians opposed to slavery. While the Royal Navy had largely suppressed the Atlantic slave trade from West Africa, trafficking in humans continued in East Africa. Abolitionists hoped that British incursions into the region might disrupt the practice by promoting alternative forms of commerce.

A master linguist, Burton gained fame as a Victorian adventurer in 1853 when, disguised as an Arab pilgrim, he gained entry into the Islamic holy cities of Medina and Mecca, which were forbidden to non-Muslims under pain of death. Shadowed by accusations that he killed an Arab boy who had uncovered his ruse, Burton lived to tell the tale in a sensational travel narrative. This triumph led Murchison to invite him to organize an RGS expedition with Speke to find the Nile's source in the great lakes of central Africa. Their first attempt ended with an attack by Somalis that left one companion dead, a javelin piercing though Burton's two cheeks, and Speke escaping capture with over a dozen stab wounds. On their second attempt, Burton reached Lake Tanganyika in 1858 too sick to continue. Partly deaf after cutting a burrowing beetle from his ear and temporarily blinded by disease, Speke pushed on to find and rename Lake Victoria. He and Burton were the first Europeans to reach either lake, but the loss of their instruments prevented a complete survey. They split over whether Lake Victoria fed the Nile and returned without finding its outlet.

Drawing large audiences to their separate RGS lectures, Burton and Speke became celebrities in Britain. Neither represented the Victorian ideal. Some listeners were disturbed by Burton's crude depictions of alleged African depravity, others by his romantic embrace of Arab culture. His graphic portrayal of native sexual practices made some shocked Victorians suspect he had participated in them. And his persistence in challenging Speke's claim to have found the Nile's source in Lake Victoria diminished the stature of both men. A follow-up RGS expedition by Speke—who seemed to make enemies easily—failed to settle the matter. "Poor Speke," Livingstone wrote, "has turned his back on the real sources of the Nile."[4] The embattled explorer died from a self-inflicted gunshot wound in 1864 while hunting on the eve of a scheduled debate with Burton at the BAAS's annual meeting, before an expected audience of thousands. The way was left open for David Livingstone.

Livingstone was made to order for Victorian tastes and soon became Murchison's most-favored explorer. He had been born in poverty, the son of a peddler, near Glasgow in 1813 and taught himself languages and science while working twelve-hour shifts as a child laborer in a cotton mill. Beaten by his pious Protestant father for reading travel and science books rather than solely religious ones, Livingstone grew to accept the Bible and science as complementary paths toward truth, and chose missionary travel as his way to serve both God and humanity. "Religion and science are not hostile," he affirmed in his first book about African travel, "but friendly to each other, fully proved and enforced."[5] Like many British evangelicals of the day, Livingstone hated the African slave trade and longed to carry civilized Christian ways to what he perceived as heathen lands. Never an effective preacher, he earned a medical degree to make himself useful on the mission field and accepted a call to the South African outback in 1840.

From his base at a mission on the northern edge of Britain's Cape Colony, Livingstone was drawn north toward people and places unknown to Europeans. Evangelizing as he went, he relied heavily on local porters and guides, with whom he often established strong bonds. Europeans avoided the interior of sub-Saharan Africa out of fear of disease and a common misperception that the region was barren desert. Livingstone had an iron constitution that seemed able to endure all manner of illness and a sustaining faith that he could find in central Africa a fertile highland, free of tropical diseases, where European trade and missions could flourish. Never lacking in ambition, he saw commerce, Christianity, and civilization as conjoined means to stop the slave trade and redeem Africa.

In 1850, reports of Livingstone's travels began filtering back to the RGS through his letters to the London Missionary Society. "Our journey across the desert was one of great labour and suffering," he wrote in one letter about a six-hundred-mile trek with two British hunters across the Kalahari Desert to the river Botetle and Lake Ngami, in present-day Botswana. Neither body of water had been seen by Europeans. The letter, read to the RGS in 1850, contained latitude readings for various points taken with a borrowed sextant and altitude estimates based on the boiling point of water. It described a land of "gigantic trees, some of them bearing fruit quite new to us," lakes and rivers swarming with "large shoals of fish," and a race of people with a "frank and manly bearing."[6] This account differed sharply from the portrait of African desolation and savagery painted by Burton and other early explorers. It was what Murchison and many empire-minded British Christians wanted to hear.

For "discovering" Lake Ngami, the RGS awarded Livingstone a twenty-five-guinea prize, which he used to buy his own sextant and surveying equipment. "The fact that the Zouga is connected with other large rivers flowing into the lake from the North," Livingstone wrote in his letter, "opens the prospect of a highway capable of being easily traversed by boats to an entirely unexplored, but, as we were told, populous region. The hopes which that prospect inspires in behalf of the benighted inhabitants might subject me to the charge of enthusiasm—a charge, by the way, I wish I deserved—for nothing good or great has ever been accomplished in the world without it."[7] Buoyed by hopes of reaching unsaved souls and unknown lands, he ventured out again in 1852.

Livingstone's fame grew out of this 1852–56 expedition, which became the subject of his best-selling 1857 book, *Missionary Travels and Researches in South Africa,* and transformed him into a near-mythic hero in Britain. On a journey chock-full of hair-raising adventures, he inscribed a T-shaped track across a vast expanse of southern Africa well known to African and Arab traders but never before visited by Europeans. First he trekked north to the Zambezi River, in what would become Britain's rich Rhodesian colonies. Then he turned west to reach the Portuguese colony of Angola on the Atlantic Coast, from where he dispatched letters detailing his discoveries to the RGS, which rewarded him with its highest award, the Queen's Gold Medal. When presenting this award, Murchison reportedly said of the still absent Livingstone, "There was more sound geography in the last sheet of foolscap which contained the results of your observation than in many imposing volumes of high pretentions."[8]

Rather than return by sea from Angola, after recovering from a near-fatal illness, Livingstone retraced his steps to the Zambezi and then followed this river east to the Portuguese colony of Mozambique on the Indian Ocean, where a British warship waited to carry him home. On this final leg of his inland travels, Livingstone became the first European to see the magnificent falls of the Zambezi, which he named Victoria in honor of his queen. "No one can image the beauty of the view from anything witnessed in England," he wrote of these falls. "It had never been seen before by European eyes; but scenes so lovely must have been gazed upon by angels in their flight."[9]

Livingstone's four-year trek took on epic proportions in the annals of European travel. Although presumably accomplished by Africans and perhaps by Arabs, no record exists of anyone crossing Africa before Livingstone: he did it one and a half times. Further, he kept detailed geographical records along the way, including regular calculations of longitude by lunar occultation, latitude by sextant, and altitude by the temperature of boiling water. "I never knew a man who, knowing scarcely anything of the method of making geographical observations, or laying down positions, became so soon as adept that he could take the complete lunar observation, and altitudes for time, within fifteen minutes," the Astronomer Royal and RGS medalist Thomas Maclear commented in 1856. "His observations of the course of the Zambezi, from Stesheke to its confluence with the Lonta, are the finest specimens of geographical observation I ever met with." These were the data demanded by mid-nineteenth-century scientific geographers and sought by the RGS. "To no one in modern times have this country and the world been more indebted for geographical knowledge and researches," an RGS report declared even before his great journey ended.[10] Livingstone's findings led to the first serviceable map of the region and filled a large blank space on the European globe.

By the time he reached Mozambique in 1856, Livingstone was sending his reports directly to Murchison, who saw them as confirmation of his theory that Africa had an elevated central basin. "*I thank you most heartily* for having made me your correspondent," he responded. "I shall indeed have the liveliest pleasure in talking over your verification of my theoretical speculations on the ancient and modern outline of Africa." In a further reply, Murchison hailed Livingstone's work as "the greatest triumph in geographical research which has been effected in our times."[11]

Best of all for Murchison, Livingstone reported finding disease-free, fertile highlands near Victoria Falls that might provide a regional base for British trade and missions. The two men shared the dream that such a colony could undermine the local slave trade, which Livingstone grimly documented in his letters. As if to claim the place for British pastoral pursuits, he planted a hundred peach and apricot pits and left them in the care of a native farmer, noting that "I have great hopes of Mosioatunya's abilities as a nurseryman."[12] Cotton, he added, could become a staple crop that would reduce British dependence on the product from the slave-owning American South. From Victoria Falls, Livingstone followed the Zambezi for most of its course to the ocean. It offered, he claimed, a navigable route from the sea to this potential British outpost. This assertion was Livingstone's greatest boast, a boon to empire and a coup for the RGS. It was also his greatest geographical blunder.

Livingstone received a hero's reception in Britain. Humble in tone yet with a flair for publicity—he wore his simple black coat and explorer's cap even for his audience with the queen—he told a story combining British aspirations for science, religion, adventure, and empire. His rise from poverty, Scottish roots, and missionary service appealed to Britons across class, ethnic, and religious lines. Even before he reached London, "in consideration of the probable interest of your work," Britain's leading publisher of science and exploration, John Murray, offered him an extraordinary contract for an account of his travels: two-thirds of the profits, a higher rate than Murray would pay Charles Darwin for *On the Origin of Species* three years later.[13] *Missionary Travels and Researches,* which Livingstone dedicated to Murchison, sold phenomenally well. The account even won over Charles Dickens, who had long satirized Victorian evangelicals as ignoring real domestic needs in favor of abstract foreign ones. In a rave review, Dickens praised the book as "a narrative of great dangers and trials, encountered in a good cause, by as honest and as courageous a man as ever lived."[14] Throughout Livingstone's stay in Britain from 1857 to 1862, throngs cheered Livingstone's every public appearance and flocked to his lectures.

For his much-anticipated first public address in Britain, Livingstone passed over the London Missionary Society, which had initially sponsored his travels for religion, for the RGS, which had commandeered them for science. "The society's rooms were crowded to excess," the *Times* of London reported of Livingstone's reception at the RGS by a "distinguished assembly" of aristocrats, government officials, military officers, and explorers that cheered virtually his every remark.[15]

Murchison, chairing the session, introduced Livingstone "as the pioneer of sound knowledge, who, by his astronomical observations, had determined the site of numerous places, hills, rivers, and lakes, nearly all hitherto unknown, while he had seized upon every opportunity of describing the physical features, climatology, and even the geological structure of the countries he had explored, and pointed out many new sources of commerce." In reply, Livingstone declared, "May I hope that the path which I have lately opened into the interior will never be shut, and that, in addition to the repression of the slave trade, there will be fresh efforts made for the development of the internal resources of the country." In such expressions, Murchison's and Livingstone's minds met and their purposes merged; they became lifelong friends and collaborators. "Both men were Scots, both were dedicated Christians, dedicated patriots and dedicated geographers," RGS historian Ian Cameron wrote, "and for the rest of his career Livingstone—like Scott a couple of generations later—became the chosen child of the Society."[16] As a self-sacrificing explorer who took extreme risks for geographical science, Livingstone became Scott's model.

Livingstone's later travels were anticlimactic. From 1858 to 1865 he led a large-scale, government-funded expedition to ascend the Zambezi by steamboat toward his proposed highlands colony. Unfortunately, on his earlier descent, Livingstone had bypassed a critical stretch of the river that included impassable rapids. His return journey foundered on these rocks. The *Times,* which had once celebrated Livingstone, now demanded his head. "We were promised trade, and there is no trade," the news voice of the empire complained in 1863. "We were promised converts to the Gospel, and not one has been made." The government ultimately recalled the expedition, but not before Livingstone's wife died of fever and the expedition's artist and ship captain both resigned. "Dr. Livingstone is unquestionably a traveler of talent," the *Times* concluded, but "it is now plain enough that his zeal and imagination much surpass his judgment."[17]

Murchison did not abandon his friend. In 1866, the RGS sent Livingstone to Africa to find the source of the Nile. Accompanied by African guides and porters, he became increasingly isolated in the Great Lakes region of central Africa. After rumors of his death reached London in 1871, the RGS dispatched a relief party. A Welsh-born journalist sent by the *New York Herald* to cover the story, Henry Morton Stanley, reached Livingstone first. Sick and out of supplies, the beleaguered explorer had sought refuge in the slave-trading town of Ujiji near Lake Tanganyika. "Dr. Livingstone, I presume?" the journalist asked in an affected British understatement. "Yes, that is my name," was the reply.[18] With supplies from Stanley, Livingstone continued his ever more desultory wanderings. He died in 1873, afflicted by multiple tropical diseases. Devoted attendants buried his heart where he died and carried his sun-dried corpse a thousand miles to the coast for transport to London, where he lay in state in the RGS map room before being buried among England's heroes in Westminster Abbey, the only explorer so honored.

Other European explorers extended the reach of Western geography during the late 1800s, but until Scott, none became such a popular hero as Livingstone. By the time of Scott's *Discovery* expedition, an explosion of imperial expansion had carried European explorers deep into every continent except Antarctica, and their sponsoring nations had colonized Africa and much of Asia. Respected but not beloved, Stanley—who returned to Africa in the service of European imperialists, confirmed Speke's claim that the Nile flowed out of Lake Victoria, and traced the Congo River from its source to the sea—pushed empire as much as discovery. "Geography," he asserted in 1884, is "a science which points with commonest inductions to . . . the paths which commerce ought to follow" and "has been and is intimately connected with the growth of the British Empire." His African work done, the RGS greeted the return of Stanley to Britain in 1890 with a grand ceremony at London's cavernous Royal Albert Hall. The society's librarian depicted the event as a celebrating "the completion of the task [the RGS] had laboured upon in setting the great features of African hydrography and orography."[19] Society members had reason to cheer. Save for Antarctica, Europeans had mapped the world.

The large blank space at the bottom of European globes had not stood out in 1850. At that time, the only distinct lines on Antarctic maps traced Ross's route along the Victoria Land coast and the Great Ice Barrier, outlined the whaling grounds of the Antarctic Peninsula and Weddell Sea, and marked a few isolated shores and islands supposedly spotted by sailors. These lines delineated less than one-tenth of the actual coast and none of the interior. But there were other gaps on European globes. The Arctic coast remained largely uncharted even though most of it was nominally under European rule. Virtually the entire interior of Africa, much of central Asia, and parts of the Americas appeared on European-drawn maps either as blank spaces or filled in with imagined features such as Ptolemy's fabled Mountains of the Moon in central Africa.

By 1900 much had changed. Late in life, the great novelist of empire Joseph Conrad recalled the profound experience of "entering laboriously in pencil the outline of [Lake] Tanganyika on my beloved old atlas, which, having been published in 1852, knew nothing, of course, of the Great Lakes. The heart of its Africa was white and big." Not only did lakes, rivers, and towns replace trackless desert on European maps of Africa, the entire continent except tiny Liberia and ancient Abyssinia became a jigsaw puzzle of European colonies, most of them British or French. Similarly, Russians pushing south from Siberia and Britons pushing north from India had filled in European maps of central Asia. The story repeated itself elsewhere. Empire drove geography and geography drove empire. As early as 1864, Prime Minister William Gladstone cautioned RGS members, "Gentlemen, you have done so much that you are like Alexander, you have no more worlds to conquer." Gladstone overlooked Antarctica. RGS historian Robert Stafford more precisely commented, "Except for the poles, the explorers had nearly worked themselves out of a job by the late 1880s."[20] For Clements Markham, this distinction made a considerable difference.

In the period after Murchison's death in 1871, Markham probably had greater practical familiarity with geography than any other Briton. He had lived it. Following in the wake of empire, by 1890, the widely traveled Markham had participated in British expeditions to the Arctic, South America, Asia, and Africa. He had also served under Murchison and others as RGS secretary from 1863 to 1888. Invited to contribute the entry on "The Progress of Geographical Discovery" to the classic 1875–89 edition of *Encyclopaedia Britannica,* Markham concluded his synopsis by declaring that along with surveying the deep-sea floor and the remote interiors of Asia, Africa, South America, and New Guinea, "The great geographical work of the present century must be the extension of discovery in the Arctic and Antarctic regions."[21]

Having been a nineteen-year-old midshipman on the Franklin searches in 1850, Markham turned first to promoting British Arctic exploration, which had faltered in the 1850s with the failure to find Franklin's lost party. The initiative passed to other nations during the 1860s and 1870s, with Norwegians first circumnavigating Spitsbergen, Austrians discovering Franz Josef Land, and Swedes first sailing the Northeast Passage. Starting in 1865, Markham led a group of self-styled RGS Arctic Associates who lobbied the government to resume exploration in the far north. "A mere quest for the Pole was not an aim which would secure influential or intelligent support," he said of this effort. "The objects of Arctic exploration, in these days, must be to obtain valuable scientific results." A cousin who joined the resulting expedition later recalled that Markham had "persistently pleaded" the cause for ten years before achieving "complete success." The approach worked. Explaining the decision to send an expedition north, Prime Minister Benjamin Disraeli specifically cited "the scientific advantages to be derived from it."[22] The pattern repeated itself three decades later for Antarctica, with Markham again pushing scientific justifications to the fore.

When the British Arctic expedition sailed in 1875, its leader, George Nares, invited Markham to join it as far as Greenland and assigned his cousin to lead its winter sledge journey toward the pole. "The Arctic Expedition achieved all that [the RGS] Council desired," Markham later claimed. "It succeeded in crossing the threshold of the unknown region, its ships attained a higher latitude than any other vessel has ever reached, they wintered further north than any human being has ever been known to have wintered before, and Captain [Albert] Markham planted the Union Jack on the most northerly point ever reached by man." Clements Markham clearly cared about these firsts and lauded the experience they afforded for naval officers, but he also hailed what he saw as the expedition's scientific results. In his list, they included "the discovery of 300 miles of new coastline," magnetic and meteorological observations at two stations, studies of tides and ice flows at high latitude, geological and biological collections, and "the discovery of a fossil forest in 82° North latitude."[23] He was putting the best possible face on an otherwise disappointing effort.

Markham gained a further chance to promote polar exploration in 1893, when he unexpectedly became the RGS president. His ascension followed a bitter row over the admission of women members that split the society. Aligned with tradition but taking no part in the dispute due to his absence on an overseas trip, Markham unknowingly emerged as a compromise candidate. He held the post for twelve years, during which time no women joined the club. In a manner recalling both the African expeditions launched by Murchison and his own Arctic endeavors, Markham focused the society's attention on the large blank space at the bottom of world maps. With no one living there and scant prospect of anyone doing so, exploring Antarctica would serve little commercial, colonial, or evangelical religious purpose, but for him geography had a higher calling. "Four main causes have led to geographical discovery and exploration," he wrote in his *Encyclopaedia Britannica* article, "namely, commercial intercourse between different countries, the operations of war, pilgrimages and missionary zeal, and in later times the pursuit of knowledge for its own sake, which is the highest of all motives."[24] The RGS's Antarctic researches fell squarely in the fourth category, with Markham having his own view of what constituted knowledge for its own sake.

In the 1890s, Markham envisioned a new Antarctic expedition yielding the same sorts of scientific results as the 1875 Arctic expedition, only more and better. Geographical mapping and surveying unexplored territory came first for him, and then magnetic, meteorological, oceanographic, geological, and biological observations. Nares had balanced all of these pursuits in his 1875–76 Arctic expedition, and Markham went so far as to call the new expedition's ship *Discovery* after Nares's main vessel. The name captured Markham's vision. He wanted to discover the earth's last uncharted large landmass and bring it under the domain of geographical science.

* * *

By the time of Scott's 1901–4 expedition, the RGS had refined its concept of geographical discovery and encapsulated it in a guidebook, *Hints to Travellers*. "The advance of geographical exploration and discovery during the last fifty years has been so rapid and continuous that there remain at the present time few parts of the Earth's surface that are entirely unknown," the 1906 edition began. "A man who only makes a hurried journey through some imperfectly known district, without proper instruments or previous training, and who is able, consequently, only to bring back with him a rough prismatic compass sketch of the route he has taken, unchecked by astronomically determined or triangulated positions, will, at the present time, find that he has not rendered any great service to geography." The book's 1901 edition declared, "The intending traveller who proposes to undertake the survey of an unexplored country should . . . have a knowledge of plane trigonometry, and those computations of practical astronomy which are necessary to enable him to fix his position in latitude and longitude."[25] Loaded with mathematical tables, equations, and instructions, *Hints to Travellers* accompanied every British Antarctic expedition. Scott carried the 1901 edition on the *Discovery;* Shackleton and Scott took the 1906 edition on their later expeditions.

Although widely used by explorers and originally coauthored in 1854 by Charles Darwin's captain on the *Beagle* expedition, Robert Fitzroy, *Hints to Travellers* took its inspiration from the growing number of Victorian tourists who wanted to join in the enterprise of geography. "One of the results of the rapid multiplication of lines of ocean steamers and of railways in distant seas and far-off countries has been to make it easy for men of comparatively brief leisure to undertake a share in exploration in the course of a vacation tour," the 1893 edition explained. "India has become commonplace, and Members of Parliament spend their holidays in Siberia, Brazil, Korea, or the Antipodes. The vacation travellers, or tourists, who are tempted by modern facilities into imperfectly-known regions are numerous, and their opportunities for collecting valuable information are great."[26]

In the British scientific tradition of amateur fact-gathering, the RGS envisioned trained travelers supplementing the findings of professional explorers in the grand mission of mapping the globe. Accordingly, in addition to updating and republishing its guidebook for nearly a century, the RGS offered classes for travelers in geographical fieldwork. The core course, a required prerequisite for all the others, was titled "Surveying and Mapping, including the fixing of positions by Astronomical Observations."[27] Preparing for their expedition, Scott and his officers took customized versions of these courses. The RGS expected the explorers to lay down accurate maps of the territory they covered even if no one was ever likely to use them for commerce, conquest, or converting the nonexistent natives.

By this time, the compulsion to map had overtaken the commercial, colonial, or evangelical purposes that mapping once served. In his *Encyclopaedia Britannica* article, Markham may have called it "the pursuit of knowledge for its own sake," but it was also for the sake of people like Markham. They wanted to know the precise location and altitude of Antarctic mountains, glaciers, and capes that they would never see and have them named for themselves and their kind. Scott and his men became their agents in this task. Of course, the *Discovery* expedition had other scientific goals, such as the magnetic survey that brought the Royal Society aboard and justified its government funding, but for Markham and the RGS, mapping was central to the mission.

The RGS equipped the *Discovery* expedition with the full range of tools then used to make topographical maps. These included sextants, theodolites, aneroids, telescopes, chronometers, compasses, plane tables, and various measuring devices—all in portable sizes for sledging. *Hints to Travellers* gave information for their use, causing Scott to panic when he lost his copy on the long Western Sledge Journey to the Polar Plateau. "The gravity of this loss," he wrote, "can scarcely be exaggerated. . . . We expected to be, as indeed were, for some weeks out of sight of landmarks. In such a case as this the sledge traveler is in precisely the same position as a ship or boat at sea: he can only obtain a knowledge of his whereabouts by observations of the sun or stars, and with the help of these observations find his latitude and longitude. To find the latitude from an observation of a heavenly body, however, it is necessary to know the declination of that body, and to find the longitude one must have not only the declination, but certain logarithmic tables. . . . Now, all these necessary data are supplied in an excellent little publication issued by the Royal Geographical Society and called 'Hints to Travellers.'" Scott was forced to rely on dead reckoning to make it out and back, but returned without a reliable record of his route.[28]

The *Discovery* expedition's two Western Sledge Journeys, the first led by Armitage in 1902 and the second by Scott in 1903, addressed one of the three principal geographical aims left over from Ross's 1839–43 expedition. Ross had sailed into the Ross Sea from the north. "He had found a high, mountainous country to the west and an ice wall to the south," Scott explained in laying out his expedition's geographical goals. "Ross was in a sailing ship and only saw things dimly and at a distance. Beyond this nothing was known. The geographical problem was, therefore, in brief, to find out what lay to the east, to the west and to the south of what Ross had seen."[29]

To the west on their Western Sledge Journeys of 1902 and 1903, the explorers found an ice sheet or Polar Plateau extending beyond the Victoria Land mountains. Scott had hoped to determine its extent by sending a party across it to the sea but fell far short of this utterly unrealistic goal. Armitage's 1902 party struggled simply to reach the plateau, and Scott's 1903 party covered only two hundred miles of its surface before turning back. "All we have done is to show the immensity of this vast plain," Scott wrote on the final day of his party's outward march. "Greenland, I remembered, would have been crossed in many places by such a track as we have made." Antarctic's ice sheet was much larger than anything in the north, he concluded. Gauging the altitude, Armitage reported it as "9,244 feet by theodolite angles, 8,727 feet by aneroids, the mean height by the two methods being 8,985 feet."[30] Scott found it 8,900 feet above sea level by aneroids, with virtually no change as he traversed it.

Assuring his expedition's geographical significance at the outset, Scott looked beyond Ross to the east before establishing winter quarters at McMurdo Sound, which both men initially described as a bay. Ross had discovered that navigable inlet between Ross Island and the Victoria Land coast, sailed deep into it, reported seeing mountains on the southern horizon, and then headed east along the Ice Barrier until he sighted land in the distance. With this, he claimed to have charted the three sides of the Ross Sea. Staying well back from the barrier in his sailing ships, Ross described it as a "solid-looking mass of ice" without "any rent or fissure" rising 150 to 200 feet from the sea. He did not know if it floated or was mostly grounded; or whether it was composed largely of compressed sea ice or glacial outflow. To him it constituted one of the world's geographical wonders, and geographers had wondered about it ever since. "Perhaps of all the problems which lay before us," Scott wrote, "we were most keenly interested in solving the mysteries of this great ice-mass."[31]

Sailing further south than Ross in McMurdo Sound in January 1902, Scott resolved that the mountains Ross reported seeing in the southern horizon were optical illusions and did not supply a backdrop for the Ice Barrier. Then he turned his attention on the barrier itself by sailing east along it. "In order that nothing important should be missed, it was arranged that the ship should continue to skirt close to the ice-cliff; that the officers of watch should repeatedly observe and record its height, and that thrice in the twenty-four hours the ship should be stopped and a sounding taken," Scott wrote.[32] They found that the Ice Barrier's sea edge extended for 490 miles.

Little now seemed as Ross had reported. "On steaming along the barrier, we soon found that Ross had exaggerated not only its height, but its uniformity," Scott noted. *Discovery* observers calculated that the height of the Ice Barrier's sea edge varied from 30 to 280 feet. "Soundings both over and under 400 fathoms. Barrier sometimes very broken and rugged in outline," Scott jotted in his diary. No one could see land at Ross's farthest east, which scotched another of his findings. The *Discovery* sailed on to where the barrier visibly terminated on solid ground, which Scott named King Edward VII Land for his monarch. Sizing up this newfound land from aboard ship, Scott noted, "With a sextant and the distance given by four-point bearing, we were able to calculate the altitude as between 2,000 and 3,000 feet."[33] Accumulating sea ice and deteriorating weather kept the explorers from landing.

Discovery instead tied up at a bight in the Ice Barrier roughly eighty miles west of Edward VII Land. Armitage took a six-person sledge party south for an overnight outing that covered about seven miles while the ship's crew inflated a tethered surveillance balloon. "The honour of being the first aëronaut to make an ascent in the Antarctic Regions, perhaps somewhat selfishly, I chose for myself," Scott confessed, "and as I swayed about in what appeared a very inadequate basket and gazed down on the rapidly diminishing figures below, I felt some doubt as to whether I had been wise in my choice." Rising eight hundred feet, Scott saw an undulating expanse of ice extending as far as he could make out with binoculars. The ice waves ran parallel to the barrier's sea edge at roughly two-mile intervals and, according to Armitage, rose and fell sharply. "Rather than crossing a series of undulations, the party had appeared to be travelling on a plain intersected by broad valleys, the general depth of which as measured by aneroid was 120 feet," he reported.[34] Although clearly an ice shelf or sheet, the explorers persisted in using the term "Ice Barrier" for the entire frozen mass. Shackleton also went up in the balloon, followed by three others before a leak ended the experiment for good. With winter coming on, *Discovery* returned to McMurdo Sound.

Sailing along the Great Ice Barrier's front could not reveal its true extent. The explorers' most revealing glimpse of this came from Scott's record-smashing Southern Sledge Journey of 1902–3. It generated the expedition's most memorable geographical findings and, more than any other event during the *Discovery* expedition, made Scott famous. It was for him what the Zambezi trip had been for Livingstone: the first public display of his mettle.

While the British scientists and scientific societies promoting the renewal of Antarctic research asserted that the *Discovery* expedition would be all about science and geography rather than a record-chasing dash to the South Pole, Markham and Scott surely had that feat in mind from the outset, even if they did not admit it. Markham later conceded that "all mention of the south pole was carefully avoided" in the official expedition instructions that he negotiated with the Royal Society. Reaching the pole could not happen at the expense of science. But the instructions did authorize Scott to determine "the nature, condition, and extent of that portion of the south polar lands" within the two Antarctic quadrants encompassing the Great Ice Barrier, and those quadrants extended to the pole. Further, Markham's initial draft instructions, which he maintained were "practically" the same as the final ones, authorized "a complete examination of the ice mass which ends in the cliffs discovered by Sir James Ross."[35]

At the time, some geographers thought the barrier ice shelf might extend to the South Pole. If it did, then (within the scope of the expedition's mandate) a small, fast-moving sledge party conceivably could get there and back in three months by racing across the Ice Barrier either on skis, the way Nansen had crossed Greenland, or by using dogs as Peary was doing in the Arctic. At least the party could set a new

Farthest South record while extending the boundaries of geographical knowledge. Scott must have discussed both options when asking the expedition's junior physician and artist, Edward Wilson, to join his Southern Sledge Party, which would depart in early November 1902. "It must consist of either two or three men in all and every dog we possessed," Wilson reported Scott saying. "Our object is to get as far south in a straight line on the Barrier ice as we can, reach the Pole if possible, or find some new land, anyhow do all we can in the time and get back to the ship by the end of January."[36] Scott and Wilson settled on Shackleton as the third man.

With these mixed goals in mind, the Southern Party left the expedition's Hut Point base on November 2 with five sledges, nineteen sled dogs, and nearly a ton of supplies, food, and instruments. A twelve-man support team left the base three days earlier to stock depots but turned back in mid-November. All were together on the thirteenth, however, when Scott's readings showed they had nearly reached the Seventy-ninth Parallel and therefore passed Borchgrevink's old Farthest South. "The announcement of that fact caused great jubilation," Scott noted in his diary. This marked about a one-degree (or seventy-mile) advance from Hut Point's latitude at 77° 47′ south. After the support team left two days later, he added, "We are already beyond the utmost limit to which man has attained: each footstep will be a fresh conquest of the great unknown. Confident in ourselves, confident in our equipment, and confident in our dog team, we can but feel elated with the prospect that is before us."[37]

Having virtually no experience on skis, driving sled dogs, or crossing an ice shelf, Scott should have felt less confidence. Almost immediately, the dogs began to fail because of tainted food and poor handling. Soon the men were doing much of the hauling, in a relay system that involved moving half of the load at a time and returning for the other half—three miles covered for each mile gained. As each dog failed, Wilson killed it and fed the carcass to the others until none of them were left. Scott gave the orders but could not bear to watch. Having packed light for quick travel, the men ran so low on food that they could think of little else. "Conversation runs constantly on food. We are all so hungry," Wilson noted while yet on the outbound journey. It grew much worse. Scurvy stalked them by the end, especially Shackleton, who ultimately could not help haul the sledges. Wilson suffered extreme snow blindness. "I never had such pain in the eye before," he noted in his diary for the day after Christmas. "It was all I could do to lie still in my sleeping bag, dropping in cocaine from time to time."[38] Throughout the ordeal, when awake in their small tent after supper or during blizzards, they often read aloud from Darwin's *Origin of Species,* the book they carried along for comfort: survival of the fittest indeed.

"If there were trials and tribulations in our daily life at this time," Scott observed, "there were also compensating circumstances whose import we fully realized. Day by day, as we journeyed on, we knew we were penetrating farther and farther into the unknown; . . . while ever before our eyes was the line which we were now drawing on the white space of the Antarctic chart." They drew this line with care—Scott mapping and measuring, Wilson sketching landscape, and Shackleton taking photographs—never forgetting their duty as agents of the RGS. As Wilson acknowledged in his diary, "Of course we must map out and sketch this new land."[39]

The party's route traced the Great Ice Barrier's western edge, where it abutted the southern Victoria Land shore. No one had seen this coastline before. With magnificent mountains, inlets, and capes, it resembled the seacoast farther north—only here it was encased in ice and devoid of life. "We are now about ten miles from the land," Scott noted on December 19. "The lower country which we see strongly resembles the coastal land far to the north; it is a fine scene of a lofty snow-cap, whose smooth rounded outline is broken by the sharper bared peaks, or by the steep disturbing fall of some valley."[40] Twice

Edward Wilson's evocative sketch of the aftermath of killing the last dog on the *Discovery* expedition's Southern Sledge Journey, January 15, 1902, from Apsley Cherry-Garrard's *Worst Journey in the World* (New York, 1922).

the men tried to reach the nearby shore to collect geological specimens, but the broken juncture between frozen land and slow-moving ice shelf left a twisted, icy ravine that they could not cross in their weakened condition.

They settled for surveying the scene and documenting their observations. Scott gauged the altitude of mountains with a theodolite and named the highest peak for Markham. He also took the angle of prominent points along the way and triangulated their relative positions. "Just about 300 miles of new coast line we have got now," Wilson noted near their Farthest South. He had captured the passing scene in detailed sketches that he later made into color paintings. "His sketches are most astonishingly accurate," Scott declared on the spot. "I have tested his proportions by actual angular measurements and found them correct."[41] From these findings, geographers learned that Victoria Land's coastal range continued south and potentially bisected the continent as part of what would become known as the Transantarctic Mountains. With the Southern Sledge Journey extending the known length of this range and the Western Sledge Journeys revealing what lay behind it, the *Discovery* expedition gave substance to the astounding idea of a vast East Antarctic Polar Plateau blanketed by a massive ice sheet of unfathomable depth.

Scott could not tell where the Ice Barrier ended in the south or how far the mountains extended because, due to insufficient food, the loss of the dogs, and the sickness of all three men, his party advanced only slightly more than 300 miles south, or about a third of the way from their ship to the pole. This mark left Scott dissatisfied. He measured the distance by sledge odometers and checked the latitude often when the sun shone through the blowing snow. "Before starting to-day I took a meridian altitude, and to my delight found the latitude to be 80°1′," Scott wrote on November 25. "All our charts of the Antarctic Regions show a plain white circle beyond the eightieth parallel. . . . It has always been our ambition to get inside that white space, and now we are there the space can no longer be a blank." But another month of sledging took the men only two more degrees south. They reached their limit on the last day of 1902. "Observations give it as between 82.16 S. and 82.17 S.," Scott noted. "If this compares poorly with our hopes and expectations on leaving the ship, it is a more favorable result than we anticipated when those hopes were first blighted by the failure of the 'dog team.'"[42] Then it was a race to the base on dwindling rations. Out and back, including relays, the party covered a total of 960 miles in three months.

Scott made the most of the record upon his return to Britain. "If we had not achieved such great results as at one time we had hoped for, we knew at least that we had striven and endured with all our might," he wrote in his account of the Southern Sledge Journey. A similar line punctuated his public lectures, which he tellingly titled "Farthest South." These lectures and his popular book, *The Voyage of the "Discovery,"* described the expedition's other exploits as well, but Scott found the public wanting records most of all. The Southern Sledge Journey supplied them in the context of an Edwardian tale of resolve in the face of adversity. The *Times* hailed it as the expedition's "most notable achievement." Markham called it "a story of heroic perseverance to obtain great results."[43] Telling it in detail took up two long chapters of Scott's twenty-chapter book. The tainted dog food from Norway provided a convenient villain. Owing to his "constitutional" weakness, as Scott termed it, Shackleton also shouldered a disproportionate part of the blame for the party's turning back. "Our invalid," Scott called him in his book.[44] Shackleton never forgot the slight and complained bitterly when Scott sent him home on the relief ship after one year while the expedition remained for two.

As news of the expedition reached Europe, geographical discovery led most accounts of its significance. "The expedition commanded by Commander Scott has been one of the most successful ever to venture into Polar regions," the *Times* declared. "He has added definitely to the map a long and continuous stretch of the coast of the supposed Antarctic continent. His sledge expeditions, south and west and east, have given us a substantial idea of the character of the interior." Although some in the Royal Society complained about the expedition remaining in place for another winter after the prescribed term of coordinated magnetic observations concluded, with one critic grumbling that "Scott's further work will be purely geographical and will not make any addition to other scientific knowledge," Markham had only praise for the venture. "Never has any polar expedition returned with so great a harvest," he stated. "The [geographical] discoveries alone were remarkable." Even before the explorers returned to Britain, the RGS gave Scott its 1904 Patron's Medal for his Southern Sledge Journey—leaving out the expedition's other accomplishments. "There was a meeting of the Royal Geographical Society and again lots of congratulations of your having changed the maps," Scott's mother wrote to her son. "I am reveling in the sunshine of your success."[45]

Shackleton, determined to redeem his honor, knew from Scott's experience that geographical discovery and record sledge journeys in adverse conditions generated fame and funds. He secured financing for his 1907 *Nimrod* venture from Scottish industrialist William Beardmore on the promise of a push to the pole in the context of a second scientific expedition to the Ross Sea basin. "I do not intend to sacrifice the scientific utility of the expedition to a mere record-breaking journey, but say frankly, all the same, that one of my great efforts will be to reach the southern geographical pole," Shackleton explained in a plea for supplemental support from the RGS.[46] Devoted to Scott, Markham opposed helping Shackleton, but under its new president, George Goldie, the RGS endorsed the venture and lent it scientific instruments. After failing to find a suitable base near King Edward VII Land, where his party could explore less familiar territory, Shackleton wintered at Ross Island's Cape Royds, just twenty-four miles north of Hut Point.

Setting out the following spring with three other men, four loaded sledges weighing about six hundred pounds each, and four Manchurian ponies, Shackleton's Polar Party passed Hut Point on November 3, 1908, and followed roughly the route that Scott had blazed six years earlier. With the ponies pulling well at first, the party traveled faster than Scott's team. But the deep snow and steady strain wore the ponies down. Collapsing one by one, they became food for the men who more and more assumed the burden of hauling the sledges. With three ponies still in harness, the party reached Scott's southern mark on November 26, more than a month quicker than their predecessors. "We celebrated the breaking of the 'farthest South' record with a four-ounce bottle of Curaçoa," Shackleton reported in his diary. "After this had been shared out into two tablespoons full each, we had a smoke and a talk before turning in. One wonders what the next month will bring forth. We ought by that time to be near our goal, all being well."[47]

The terrain from this point was new, and the Polar Party began mapping it in earnest using a theodolite and compass. "The land now appears more to the east, bearing south-east by south, and some very high mountains a long way off," Shackleton observed on November 28. The pole clearly lay due south behind those ramparts. Expedition cartographer and physician Eric Marshall "is making a careful survey of all the principal heights," Shackleton added. "He does this regularly."[48]

To their relief, the explorers soon discovered an extraordinarily broad glacial slope—the widest yet seen—ascending through the mountains toward the south. Named Beardmore Glacier for the expedition's patron, it became the party's route to the plateau. "The pass through which we have come," Shackleton wrote, "is flanked by great granite pillars at least 2000 ft. in height and making a magnificent entrance to the 'Highway to the South.' It is all so interesting and everything is on such a vast scale that one cannot describe it well. We four are seeing these great designs and the play of nature in her grandest moods for the first time." Large patches of sheer ice and deep crevasses complicated their ascent, with one snow-covered chasm claiming their last surviving pony, Socks. The long-suffering animal was near collapse anyway, but its abrupt disappearance into what Shackleton called "a black bottomless pit" deprived the men of needed food. Soon they faced the same problem of dwindling rations that had hindered Scott's Southern Sledge Journey. By December 22, Shackleton prayed, "Immediately behind us lies a broken sea of pressure ice. Please God, ahead of us there is a clear road to the Pole."[49] There wasn't.

After twenty-four days of mostly man-hauling sledges up the glacier, the Polar Party finally reached the plateau. "It is a wonderful thing to be over 10,000 ft. up at the end of the world almost," Shackleton noted on December 28, but soon added, "Physical effort is always trying at a high altitude." Heading south on the Polar Plateau, the party fought a constant headwind and severe altitude sickness. "The sensation is as though the nerves were being twisted up with a corkscrew and then pulled out," Shackleton complained. "The Pole is hard to get." Reaching it on the food available would require the party to average eighteen miles per day across the plateau, but even by reducing their load with regular depots, they never approached that mark. "The main thing against us," Shackleton wrote on January 4, "is the altitude of 11,200 ft. and the biting wind. Our faces are cut, and our feet and hands are always on the verge of frostbite."[50] He later explained, "We had weakened from the combined effect of short food, low temperature, high altitude and heavy work."[51]

Two days of hurricane-force winds and temperatures down to −40°F in early January confined them to their tent and sealed their fate. They halted on January 9 at 88° 23′ south—112 statute miles from the pole—having covered over 1,000 miles (including relays) in ten weeks and beaten Scott's record by 430 miles. As precise as their final latitude sounds, Shackleton calculated it by dead reckoning. By this point, the party had left its theodolite behind and lost its sledgemeter. All four men swore by it, however, and few besides an embittered Markham openly doubted their word. "While the Union Jack blew out stiffly in the icy gale that cut us to the bone, we looked south with our powerful glasses, but could see nothing but the dead white snow plain. There was no break in the plateau as it extended towards the Pole, and we feel sure that the goal we have failed to reach lies on this plain," Shackleton declared. "Whatever regrets may be, we have done our best." He later said to his wife, "A live donkey is better than a dead lion, isn't it?" She replied, "Yes, darling, as far as I am concerned."[52]

Racing from depot to depot with no margin for error, the Polar Party made the seven hundred miles back to base in seven weeks with a story of perseverance in adversity that matched the tale of Scott's Farthest South. "And no man told a story better," one editor said of Shackleton, who had advance contracts for his narrative with newspaper, magazine, and book publishers.[53] Largely written on the voyage home with the aid of a New Zealand journalist and featuring a balance of geographical discovery and daring adventure, wired accounts from Shackleton began appearing even before the explorer reached London. Press reports proclaimed him the "lion" of the season despite his failure to reach the pole.[54] He was knighted by the king and, like only Stanley, Nansen, and Scott before him, was received by the RGS at the Royal Albert Hall, where for the first time motion pictures taken on an expedition illustrated the lecture. To mark the occasion, the society issued a new Antarctic map incorporating the expedition's geographical survey and showing landmarks that Shackleton named for such RGS officials as Goldie, Mill, Keltie, and Leonard Darwin. Even if no one had yet reached the pole, the Polar Party had resolved its geography. The RGS map showed that there were no fixed features on the vast East Antarctic ice sheet—only a spot yet to be reached.

While in the Antarctic, the *Nimrod* expedition completed two more notable bits of geographical research. During the Northern Sledge Journey across the Ross Sea ice from Cape Royds toward the

South Magnetic Pole, Douglas Mawson used triangulation and traverses to survey the Victoria Land coast and mountains with a precision not previously matched from ships. His findings appeared on the new RGS map. Finally, on leaving the Ross Sea, *Nimrod* steamed west along the Antarctica coast to a point somewhat farther than earlier expeditions. "On the morning of March 8," Shackleton noted, "we saw, beyond Cape North, a new coast-line extending first to the southwards and then to the west for a distance of over forty-five miles. We took angles and bearings. . . . We would all have been glad of an opportunity to explore the coast thoroughly, but it was out of the question; the ice was getting thicker all the time, and it was becoming imperative that we should escape to clear water without further delay."[55] Faulty sightings by an 1838–42 American navy expedition commanded by Charles Wilkes suggested that this shoreline turned sharply northwest. Reports from the *Nimrod* all but erased the old Wilkes Land coast from new charts. Further evidence on this and other points of Antarctic geography would have to wait for Scott's return.

In the society's 1874 obituary for David Livingstone, which credited the explorer with opening "nearly one million square miles of new country, equal to one-fourth the area of Europe," RGS President Bartle Frere commented, "It will be long ere any one man will be able to open so large an extent of unknown land."[56] Of course, that land was fully inhabited, and Livingstone opened it to Western geography with local help. As Markham and others had by then begun to stress, two much vaster areas remained virtually unknown to all humans: the deep-sea floor and Antarctica. By tracking the Transantarctic Mountains and crossing to the Polar Plateau, *Discovery*'s Western Sledge Journeys and *Nimrod*'s thrusts toward the magnetic and geographical poles opened over three million square miles of East Antarctica. Yet the deep sea had come first. In an epic 1872–76 voyage that served as a model for the *Discovery* expedition, naturalists aboard HMS *Challenger* opened the oceans to scientific study. By revealing the impact of the Antarctic on marine biology and the global climate, the *Challenger* naturalists helped to launch the heroic age of Antarctic science.

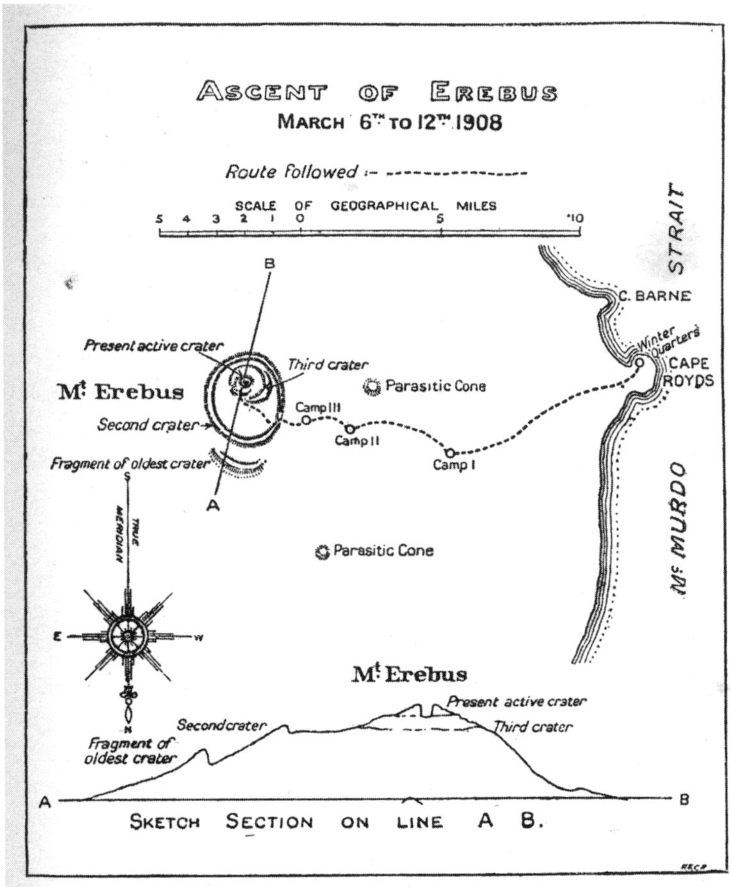

CRATER OF MOUNT EREBUS AND SECTION

Map tracing the first ascent of Mount Erebus (1908),
from Ernest Shackleton's *Heart of the Antarctic* (Philadelphia, 1909).

CHAPTER 4

In *Challenger's* Wake

"THE SEA COVERS NEARLY THREE-FOURTHS OF THE surface of the earth, and, until within the last few years, very little was known with anything like certainty about its depths," the Scottish naturalist Wyville Thomson wrote in 1873. "The popular notion was, that after arriving at a certain depth the conditions became so peculiar . . . as to make life of any kind impossible, and to throw insuperable difficulties in the way of any attempt at investigation. Even men of science seemed to share this idea." Because no one saw any use for the oceanic depths, Thomson explained, everyone ignored them. Fishing, whaling, and sailing took place at or near the surface, where life flourished. According to Edward Forbes, Thomson's predecessor as the University of Edinburgh's professor of natural history, water pressure became too great, sunlight too scant, and oxygen too scarce below three hundred fathoms (or eighteen hundred feet) for life to exist. After Forbes's death in 1854, Thomson came to realize that scientists had never actually looked for life on the ocean floor and had closed their eyes to evidence of its existence.[1]

In the late 1850s, the triumph of electric telegraphy in revolutionizing communications on land led ambitious European and American entrepreneurs to try connecting their continents by a telegraph cable from Ireland to Newfoundland. "Practical men mapped out the bed of the North Atlantic," Thomson noted, "and devised ingenious methods of ascertaining the nature of the material covering the bottom."[2] The influential physicist William Thomson, later made Lord Kelvin, invested heavily in the project and worked to bring the Royal Society on board with the help of the society's vice president, invertebrate zoologist and science popularizer William B. Carpenter.

The project required deep-sea sounding, dredging, and trawling. As early cables snapped or failed, engineers hauling them up from miles down found them entangled or encrusted with living things. When the effort to lay a transatlantic cable finally succeeded in 1866, attention turned toward other oceans. Soon the globe was girdled with electric cables. The deep-sea floor had become useful for commerce, and so scientists began studying it.

With the Royal Society's backing, Wyville Thomson, Carpenter, and the wealthy Welsh conchologist J. Gwyn Jeffreys conducted deep-sea dredging expeditions aboard the British navy vessels *Lightning* and *Porcupine* each summer from 1868 to 1870. The Admiralty did not specially outfit these ships for research, and the vessels did not venture more than a couple of weeks' cruising distance from Britain in the brief time afforded for each trip. At various stations in the North Atlantic and Mediterranean Sea, researchers sounded the bottom, collected samples, and measured the water temperature at different levels. Much of the equipment was experimental. Nevertheless, the findings astounded experts and intrigued the public. By looking where no scientist had looked before, Thomson and his colleagues discovered a new world beneath the sea.

Four findings from these voyages pointed researchers back in geologic time and south toward Antarctica. First, the seabed had ridges, valleys, and plains like surface land—only much grander in scale. Second, white ooze composed largely of the chalky remains of tiny plankton covered much of the sea floor. "One can have no doubt, on examining this sediment, that it is formed in the main by the accumulation and disintegration of the shells of globigerina," Thomson wrote, "an animal formation in fact being formed very much in the same way as in the accumulation of vegetable matter in a peat bog, by life and growth above, and death, retarded decomposition, and compression beneath." Within this globigerina ooze, Thomson tentatively detected a gelatinous matter resembling egg whites that Royal Society Secretary T. H. Huxley hastily christened *Bathybius haeckelii* in honor of German evolutionist Ernst Haeckel, who had speculated about the origins of life in marine slime. "If this have a claim to be recognized as a distinct living entity," Thomson commented on the supposed substance, "it must be referred to the simplest division of the shell-less rhizopeda, or if we adopt the class proposed by Professor Haeckel, to the monera."[3] Some evolutionists dared to hope it might represent the primordial source of all life.

A third finding also excited widespread public interest. Life on the bottom was not limited to simple forms. The dredges hauled up sponges, corals, crinoids, starfish, sea urchins, crustaceans, mollusks, and even some true fishes, many in forms never before seen by humans except perhaps as fossils from past geologic periods. A few ancient species, thought extinct, were found alive. "These abysses of the ocean," Thomson wrote in his account of the expeditions, "are inhabited by a special deep-sea fauna, possibly as persistent in its general features as the abysses themselves." Further linking the modern abyss with the geologic past, the composition and structure of the white ooze resembled that of the ancient chalk beds of Europe. "We might be regarded," Thomson proclaimed, "in a certain sense as still living in the cretaceous period."[4]

Finally, the researchers found that, unlike near the surface, temperature in the deep did not drop at higher latitudes or greater depths. Even major surface currents like the Gulf Stream had little impact below two hundred fathoms. Instead, they detected a pattern of distinct temperature layers spanning the entire survey area in the

North Atlantic. "The sea was warm or cold at all depths, according to the source from which each layer or current of water was derived," Thomson wrote.[5]

Deep in the North Atlantic, Thomson and Carpenter discerned a slow, steady undercurrent of cold water, only a few degrees above freezing, flowing from south to north. It must come from polar regions, they reasoned, and because cold water is heaver than warm water, gravity could propel it along the seabed. While large landmasses restrict "the communication between the North Atlantic and the Arctic Sea," Thomson noted, "the communication with the Antarctic basin is as open as the day;—a continuous and wide valley upwards of 2,000 fathoms in depth stretching northward along the western coasts of Africa and Europe."[6] Cold surface water, they proposed, sinks in the Antarctic, travels north along the seabed, gradually warms, and rises in the far north. Acting like the coolant in a refrigeration system, an Antarctic undercurrent could maintain the ocean depths in a steady state while slowly circulating seawater from top to bottom and south to north.

For Victorian scientists struggling to come to terms with Darwinism, finding layers of constant temperature in the deep oceans had profound significance: it could account for the distribution of sea creatures across the globe. At depth, the tropics would not serve as a natural barrier between temperate or even polar zones. It might also explain the discovery of ancient species in deep oceans. If environmental change drove organic evolution, Thomson speculated, older forms might survive in a deep-sea environment stabilized by an Antarctic undercurrent. Zoologists could look there and toward Antarctica itself for so-called living fossils and perhaps find proof of Darwin's still-novel theory.

These findings from the summer cruises of *Lightning* and *Porcupine,* coupled with interest in the deep seas generated by transoceanic telegraph cables, stirred the British government to launch the *Challenger* expedition and point it toward the Antarctic. Once proposed, the venture came together rapidly. Late in 1871, the Royal Society appointed a committee composed of Thomson, Carpenter, Jeffreys, Huxley, Joseph Hooker, the future Lord Kelvin, and others to consider Carpenter's call for a government-funded deep-sea dredging expedition to circumnavigate the globe. The committee promptly recommended that the navy outfit a ship, with a staff of civilian scientists on board, "to investigate the *Physical Conditions* of the *Deep Seas* in the great ocean-basins—the North and South Atlantic, the North and South Pacific, and the Southern Ocean (as far as the neighborhood of the great ice-barrier)." The expedition would also "examine the distribution of *Organic Life* throughout the areas traversed" with particular reference "to the connection of the present with the past condition of the globe."[7] Scarcely three months after the society transmitted this recommendation to the government, the Admiralty accepted it and rushed to launch the expedition by the end of 1872.

"An early day of the ensuing week will witness the departure of the most important naval surveying Expedition which has ever been sent forth by any country," the *Times* of London reported on December 5, 1872.[8] HMS *Challenger,* a square-rigged corvette with auxiliary steam power refitted as a floating marine laboratory, became the first government ship ever specially dispatched for oceanographic research. Despite his young family, the forty-two-year-old Thomson agreed to lead the scientific staff on the anticipated four-year cruise. Five junior researchers accompanied him, including thirty-one-year-old Canadian-born naturalist John Murray. The original Royal Society committee, augmented by the addition of evolutionary biogeographer Alfred Russel Wallace, drafted the expedition's instructions.

"The principal object of the proposed expedition is understood to be to investigate the physical and biological conditions of the great ocean basins," the instructions began. Scarcely mentioning telegraph cables, they dwelled instead on the scientific benefits of deep-sea research: "The investigation of various problems relating to the past history of the globe, its geography at different geological epochs, and the existing distribution of animals and plants, as well as the nature and causes of oceanic circulation, will be greatly aided by a more accurate knowledge of the ocean-bed."[9] In addition to sounding, dredging, and measuring deepwater temperatures at regular stations, the expedition was to collect biological specimens and keep regular records of air pressure and temperature. Since winds blow toward areas of relative low air pressure, finding constant low pressure in the far south would fit Thomson's hypothesis that water carried north by the Antarctic undercurrent returned south as moisture in the air.

The expedition's route reflected its mission. After crisscrossing the Atlantic four times and rounding the Cape of Good Hope, *Challenger* was to dip south across the Indian Ocean by way of the Kerguelen Islands to the Antarctic ice pack, then take a circuitous path through the Pacific before recrossing the Atlantic to return home. "This route will give an opportunity of examining many of the principal ocean phenomena," the instructions noted, "including . . . the specially interesting fauna of the Antarctic Sea."[10] In an era when scientists had just begun reinterpreting nature from an evolutionary perspective, a comparison between this little-known fauna and the better-known one of the Arctic Ocean intrigued zoologists, climatologists, and oceanographers looking for links over time and space.

The committee wanted *Challenger* to approach the ice pack from the Indian Ocean because it presented Antarctica's least familiar face. Earlier explorers had discovered large Antarctic bays—the Ross and Weddell seas—south of the Pacific and Atlantic oceans. If this pattern held in the Indian Ocean, it might provide clues for understanding ocean circulation and the Antarctic undercurrent. As these instructions suggested, the expedition's organizers looked toward Antarctica for answers to basic scientific questions.

* * *

The *Challenger* naturalists found answers—and new questions—everywhere they looked, especially in the Antarctic. Along the way, they sent back observations and preliminary findings to the Admiralty, the Royal Society, the *Times* of London, and the science journal *Nature*. The expedition traveled eighty thousand statute miles between December 1872 and May 1876, spending about half of this time at sea. "In consequence of the special nature of the mission," expedition naturalist Henry Moseley complained, "the sea voyages were tedious and protracted, the ship being constantly stopped on its course to sound and dredge."[11] Thomson reported stops at 362 observing stations. The data overwhelmed him. He collapsed in 1881 and died the following year, after writing two volumes on the Atlantic portion of the voyage. Murray took over the task of preparing the expedition's technical reports, which ran to fifty volumes and gave birth to the modern science of oceanography. Both men were knighted for their service.

Although the results fascinated scientists and captured widespread public attention, the work at sea was deadly dull. The deepest station sounded at about five miles below the surface, the average at nearly three miles. At such depths, it could take the ship's officers and crew twelve hours or more to lower and raise the equipment. "At first, when the dredge came up, every man and boy in the ship who could possibly slip away, crowded round it, to see what had been fished up," Moseley recalled. "Gradually, as the novelty of the thing wore off, the crowd became smaller . . . and as the same tedious animals kept appearing from the depths in all parts of the world, the ardour of the scientific staff even abated." Finger-pointing invariably ensued when the dredge came up empty or snagged on the bottom. Over a quarter of the crew deserted, many in Cape Town or Melbourne just before or after the strenuous Antarctic portion of the trip. Noting that conditions could have become much worse, T. H. Huxley toasted

Thomson on his return, "When men are shut up together in a limited society, whether it be a cathedral town or a ship, they begin to hate one another unless the bishop is a very wise person. In this case I do not doubt that the bishop was a very wise person."[12]

Anticipating difficulties during the unusual mission, the Admiralty took special care in picking the crew and outfitting the ship. All the sailors, officers, and scientists volunteered for the voyage. For captain, the navy tapped George Nares, an exceptional surveyor who had served in the Arctic during the Franklin searches. Recalled to lead the 1875 British Arctic Expedition after completing *Challenger*'s work in the Antarctic, Nares was replaced by Frank Thomson for the final eighteen months of the cruise. Although the captain retained overall command, the civilian scientists called the shots on matters of where, when, and how to conduct research—a precedent that the Royal Society later tried to impose on the *Discovery* expedition. Wyville Thomson and the captain had paired cabins with a shared sitting room, and all on board had larger quarters than normal for a navy vessel. "Sixteen of the eighteen 68-pounders which form the armament of the *Challenger* have been removed, and the main-deck is almost entirely set aside for the scientific work," Thomson reported. "I feel justified in going even so far as to say that the arrangements for scientific work in the *Challenger* leave little or nothing to be wished for."[13] Navy discipline was relaxed to facilitate research.

"Never had any expedition yet left our shores so well provided with the means of adding to the hydrographical and scientific knowledge of the world," Nares declared on the ship's departure.[14] Although he stressed the practical value of potential findings for laying telegraph cables, others could not suppress their sheer wonder at what lay beneath the sea. One sailor penned doggerel about the scientists and their mission:

> Don't yer see these learned bosses have come to search the ocean,
> But for what, old son, 'twixt you and I, I'm blow'ed if I've a notion.
> I've 'eard 'em talk of Artic drift and valleys under water,

> And specs next week to find they've nab'd old Davy and his darter.
> Of course you know they've got to find the link atween the species,
> Some say as there's a 'boon aboard as liks it all to pieces.[15]

After reviewing expedition plans and preparations, RGS President Francis Galton predicted, "The voyage of the *Challenger* would be different from any other that had ever been undertaken by this or any other country." He then added about its backers, "Of course, they might reasonably look for great results."[16]

They were not disappointed. "The scientific results of the *Challenger* expedition have far exceeded our most sanguine anticipation," Royal Society President Joseph Hooker declared even before the cruise ended.[17] Like many British naturalists, he had been following the expedition's progress closely from letters and reports sent back by Thomson and others. In general, those results refined the four principal findings of earlier dredging expeditions.

Through soundings conducted during the voyage, Thomson extended his rough map of sea-floor topography to encompass the globe. "We now know," he declared, "that the sea covers a vast region which is to a certain degree comparable with the land—a region which has its hills, valleys, and great undulating plains." The expedition found its deepest bottom at 4,475 fathoms in the western Pacific, or 2,000 fathoms deeper than anything found from *Lightning* or *Porcupine*. "The thermometers, which were tested to withstand a pressure of four tons on the square inch, were all broken with the exception of one," the *Times* reported on this deepest sounding. The water at depth was so still that some mercury from the broken thermometers dropped straight down into the sounding tin below.[18]

Deeper soundings enabled the *Challenger* naturalists to refine emerging scientific views of the ocean floor. The white ooze found on the deep-sea bed by earlier expeditions gave way to a transitional gray ooze at about 1,500 fathoms and then, at about 2,200 fathoms, to a red clay bottom. This finding led Thomson to accept Murray's explanation that the microorganisms whose shells created the white ooze lived at or near the surface and that, by a certain depth, acidic deep-ocean water dissolved calcite from the remains raining down below. "The red clay is essentially the insoluble residue," Thomson reported from Australia, "of the calcareous organisms which form the Globigerina ooze after the calcareous matter has been by some means removed." Anywhere they sounded, those on board soon could predict the composition of the bottom simply from the water's depth. Nowhere, however, did they find the gelatinous substance supposedly detected by earlier expeditions. Thomson suggested that it resulted from dousing deep-sea specimens with preserving fluid. "Prof. Thomson speaks very guardedly," Huxley conceded in *Nature,* "but since I am mainly responsible for the mistake, if it be one, of introducing this singular substance into the list of living things, I think I shall err on the right side in attaching greater weight than he does to the view which he suggests."[19]

As with earlier expeditions, discoveries about deep-sea creatures sparked the most interest. "The Challenger Expedition is the final establishment of the fact that the distribution of living beings has no depth-limit; but that animals of all the marine invertebrate classes . . . exist over the whole of the floor of the ocean," the *Times* of London declared. "Never since King John proclaimed at Hastings his dominion over the oceans have our ships brought home more precious spoil."[20] Many deep-sea specimens caught in *Challenger*'s dredge had enlarged eyes; others, none at all. Some had a phosphorescent glow, apparently to compensate for the lack of sunlight, or strikingly bizarre forms. Most represented species new to science, with many of these collected at the expedition's most southerly stops. "The deep-sea fauna of the Antarctic has been shown by the *Challenger* to be exceptionally rich, a much larger number of species having been obtained [there] than in any other region visited," Murray reported in the scientific manual for Scott's *Discovery* voyage. Noting the prevalence of large mouths and sharp spines on the creatures, the manual added, "The depths of the sea bear evidence to the truth of the Frenchman's summing up of Life as the conjunction of the verb 'I eat,' together with its terrible correlative 'I am eaten.'"[21]

From his earlier work, Thomson had predicted that the same deepwater species would be widely distributed across the ocean floor and that many would be living fossils from the Cretaceous period. *Challenger's* haul upset his thinking. "The fauna of the deep sea is wonderfully uniform throughout," Thomson told the BAAS upon his return, "and yet, although in different localities the species are evidently representatives, to a critical eye they are certainly *not* identical." Further, although "the general character of the assemblage of animals is much more nearly allied to the cretaceous than to any recent fauna," current species "are certainly in *very few* instances identical" with those of the Cretaceous period.[22]

Deepwater species evolved both over time and as they migrated, Thomson concluded, with the evolutionary pattern pointing back toward the wide moat of water surrounding Antarctica. "The most characteristic abyssal forms, and those which are most nearly related to extinct types, seem to occur in greatest abundance and of largest size in the Southern Ocean," he wrote, "and the general character of the faunae of the Atlantic and the Pacific gives the impression that the migration of species has taken place in a northerly direction."[23] This, of course, nicely fit his theory of an Antarctic undercurrent.

From observations aboard *Lightning* and *Porcupine,* Thomson had argued that cold waters of the deep Atlantic came from Antarctica. Water-temperature readings taken during the *Challenger* expedition confirmed this theory and extended it to encompass the Pacific and Indian oceans. "Like the similar mass of cold bottom-water in the Atlantic, the bottom-water of the Pacific is an extremely slow indraught from the Southern Sea," he reported from Chile. "That it is moving, and moving from a cold source, is evident from the fact that it is much colder than the mean winter temperature of the area which it occupies, and colder than the mean temperature of the crust of the Earth; that it is moving in one mass from the southwards is shown by . . . the gradual rise of the bottom-temperatures to the northward." Reflecting this view, Thomson began referring to the world's three great oceans as mere "gulfs" of the Southern Sea.[24] With this, the *Challenger* expedition turned the world upside down from how Europeans traditionally saw it and moved Antarctica from the periphery of oceanic circulation to the center. As scientists came to understand the process, cold water flowing from Antarctic sources across the seabed to the Northern Hemisphere cools the ocean depths and regulates the global climate.

The theory of an Antarctic undercurrent rested on hundreds of water temperature readings taken across the globe. In contrast, Thomson's elaborate but less authoritative account of how Antarctica fit in the process relied largely on circumstantial evidence collected during the three weeks in February 1874 that *Challenger* spent among Antarctic icebergs and skirting the ice pack, never penetrating more than eight nautical miles south of the Antarctic Circle. With summer drawing to a close and his ship not reinforced for sailing in ice, Nares feared that a longer stay or deeper intrusion would doom the expedition. As it was, it suffered a narrow escape when it collided with an iceberg during a gale, losing its jibboom and nearly toppling its topgallant masts.

During this period the crew sounded, dredged, and took deep-water readings at four stations among the icebergs, recording depths ranging from eight thousand to twelve thousand feet. *Challenger* approached the pack south of the Kerguelen Islands in the Indian Ocean, where Thomson thought a bay like the Ross or Weddell seas should exist. "At first, all the icebergs seen were numbered each day, and their positions noted down," Moseley wrote, "but when we came to have 40 in sight at once this plan was abandoned." The dredge brought up fragments of feldspar, quartz, granite, slate, gneiss, and other continental rocks and minerals. Because this material clearly came from icebergs calved from Antarctic glaciers, the naturalists concluded that Antarctica must be a continental landmass. Ross and other explorers had reported land in the region but never knew if what they saw represented oceanic islands or parts of a continent. Based on the dredging, Thomson confidently proclaimed "that the area within the parallel of 70° S. is continuously solid, that is to say, that it is either continuous land or dismembered land fused into the continental form by a continuous ice-sheet."[25]

As this comment suggested, dredging did not resolve whether Antarctica was contiguous land or an archipelago of continental islands. Available evidence pointed toward the latter. "The icebergs . . . were usually from a quarter to half a mile in diameter, and about 200 feet high," Nares wrote. "They were remarkably clear of rocks or stones, although, each time we have dredged, sufficient evidence was brought up to show that the bottom of the sea is fairly paved with the debris brought by them from Antarctic lands. In shape, they were always tabular." One man on board compared them to "sugar-covered cakes divested of all colored ornaments."[26] The *Challenger* naturalists thought these distinctive flat bergs must come from flat land, unlike the familiar pinnacled icebergs of the North Atlantic that calved from Greenland's mountain glaciers.

This observation, combined with Ross's description of the Great Ice Barrier, led the naturalists to infer that Antarctica consisted mostly of low islands and shoals covered by an ice sheet that stood less than two thousand feet high. As snow accumulated on the ice sheet, they proposed, the added weight would push its sides outward across the low-lying area and into the surrounding ocean, creating the Ice Barrier that Ross described as rising two hundred feet above the sea. Like an iceberg, the floating end of a two-thousand-foot-high glacier should rise nearly two hundred feet above sea level. As the ice shelf extended further from the support of land, they reasoned, chunks would break off as tabular icebergs. Drifting north with the current, they would drop continental debris from their undersides and feed ice-cold melt-water into the Antarctic undercurrent. Even the Ross and Weddell seas, Thomson added, might be "notches" in the Ice Barrier caused by warm water from the great oceans to the north. "The 'continent,'" he suggested, "may consist to so great an extent of ice as to be liable to have its outline affected by warm currents. . . . If this be so it would at all events indicate that the 'Antarctic continent' does not extend nearly so far from the Pole as it has been supposed to do."[27]

To complete his account of ocean circulation, Thomson drew on the expedition's finding of low air pressure, strong west winds, and stormy weather in the Southern Ocean. He and Carpenter speculated that water flowing north in the Antarctic undercurrent might return south as airborne water vapor in a vast atmospheric cyclone characterized by the inward, clockwise circulation of west and northwest winds around a polar low-pressure center. The Northern Hemisphere had more surface land than the Southern Hemisphere, they noted, and would retain more solar heat. This would cause greater evaporation in the north and greater precipitation in the south. "One thing we know," Thomson baldly asserted about the entire Antarctic region, "precipitation throughout the area is very great, and that it is always in the form of snow."[28] Low pressure drew moist air south, he proposed, causing heavy snowfall on the Antarctic ice sheet, which led to the icebergs that fed the deep-sea undercurrent and completed the cycle of ocean circulation. Of course, *Challenger* never got closer than sixteen hundred miles from the South Pole. If it had, Thomson might have found that high air pressure and low precipitation typify the mainland climate.

Thomson and Carpenter had devised their doctrine of ocean circulation from their earlier research in the North Atlantic. They hoped that the *Challenger* expedition would provide proof. On receiving news of its findings in the Southern Ocean, Carpenter boasted, "The observations upon ocean-temperature . . . not only prove comfortable in every particular to the doctrine they were designed to test, but do not seem capable of any other explanation." To one persistent critic, Scottish climatologist James Croll, he sneered in the journal *Nature,* "As it is usually considered in scientific inquiry that the verification of a prediction affords cogent evidence of the validity of the hypothesis on which it is based, I venture to submit that so far my case has been made good."[29] Yet the speculative nature of these claims plainly cried out for data from the Antarctic mainland, which naturalists on later expeditions sought to supply.

Although wrong in many respects about Antarctica's coastline, topography, and climate, Thomson succeeded in turning the *Challenger* expedition's scattered findings into the first comprehensive scientific description of the South Polar region. After Thomson's death in 1882, Murray became the steward of *Challenger* science and a champion of polar research. He accepted much of Thomson's thinking about Antarctica, and his advocacy helped to make it widely accepted. In a 1905 book on Antarctic exploration, RGS Librarian Hugh Robert Mill wrote of *Challenger's* voyage, "It was scarcely an Antarctic expedition, yet more real knowledge of the nature of the Antarctic regions was obtained in the course of it than in any other voyage up to that time, Ross's excepted."[30] Its findings profoundly influenced the next generation of Antarctic researchers.

When Clements Markham became RGS president in 1893 and made an Antarctic expedition his top priority, he called on Murray to address the society on the scientific justification for the venture.

Although his well-publicized lecture covered terrestrial magnetism, geographical discovery, and other issues, Murray stressed the prospects for major discoveries in oceanography and meteorology. He saw these two fields on the cusp of breakthroughs in the understanding of global water and air circulation that would put Antarctica at the center of how science understood the sea and sky.

By this time, Murray had examined enough records from the *Challenger* expedition to appreciate (if not fully understand) the complexity of ocean currents and deepwater temperature patterns. Attempting to make sense of this data had driven Thomson toward an early grave and generated calls from researchers for more information from more places, especially the Antarctic. Existing data showed the deep seas divided like a layer cake, with warm and cold currents circulating though a global system linked mainly in the south. Cold, deep water carried north by the Antarctic undercurrent remained a critical element in the pattern, but its primary return route now seemed more likely to pass through sea than air.

Sorting out ocean circulation required more evidence from the Antarctic, Murray told the RGS in a part of his address that seemed designed to impress listeners with the matter's complexity. "The general results of all the sea-temperature observations by Cook, Wilkes, Ross, and the *Challenger,* in the Antarctic Ocean shows that a layer of cold water underlies in summer a thin warm surface stratum and overlies another warm but deeper stratum," he noted. "The cold water found at the greater depths of the ocean probably leaves the surface and sinks toward the bottom in the Southern Ocean between the latitudes of 45° and 56° S. These deeper, but not necessarily bottom, layers are then drawn slowly northward toward the tropics . . . [while the overlying] layers of relatively warm water appear likewise to be slowly drawn southwards to the Antarctic." To discover the role of polar waters in this global pattern, Murray said, "a fuller examination of these waters is most desirable at different seasons of the year, with improved thermometers and other instruments. Here, again, a new Antarctic expedition would supply the knowledge essential to a correct solution of many problems in Oceanography."[31]

Murray spoke in similar terms about the need for data on polar weather. "Our knowledge of the meteorology of the Antarctic regions is limited to a few observations during the summer months in very restricted localities," he stated, "and is therefore most imperfect." The *Challenger* expedition found low air pressure, westerly winds, and high rainfall in the ocean latitudes of the Roaring Forties and Furious Fifties, but readings farther south hinted that winds there blew in the other direction. "All the teaching of meteorology therefore indicates that a large anticyclone with a higher pressure than prevails over the open ocean to the northwards overspreads the Antarctic continent," Murray stated, in support of the view that Antarctic winds spread outward in a counterclockwise, easterly or southeasterly direction from the pole. "It is probable that about 74° S. the belt of excessive precipitation has been passed, and it is even conceivable that at the pole precipitation might be very little in excess of, or indeed not more than equal to, evaporation. Even one year's observations at two points on the Antarctic continent might settle this point." The matter was critical for understanding the world's climate, Murray concluded. "It is impossible to over-estimate the value of Antarctic observations for the right understanding of the general meteorology of the globe."[32]

To investigate the issues, Murray sought an Antarctic expedition modeled on the *Challenger* effort, with two navy vessels ferrying scientists to research sites. He envisioned shipboard researchers plumbing the ocean depths while two shore-based parties would winter on the Antarctic mainland conducting meteorological, magnetic, and other research. "The wintering parties," he proposed, "might largely be composed of civilians, and one or two civilians might be attached to each ship; this plan worked admirably during the *Challenger* expedition." The Royal Society placed greater emphasis on magnetic research than Murray, but it largely embraced his view of the project and tapped him as one of its representatives on the committee planning the *Discovery* expedition. The RGS's Markham, in contrast, though he used Murray and the Royal Society to gain government support, saw the venture as a navy exercise modeled on the voyages of Cook, Ross, and Franklin, in which scientists were subordinate to navy officers. As early as 1899, he warned a fellow RGS member involved with planning the expedition, "Do not let that overbearing fellow Murray meddle more than is necessary. Of course he cannot have things like the 'Challenger.'" This conceptual difference set the stage for a clash of wills in 1901, after the government agreed to provide partial funding and navy personnel for an Antarctic expedition planned by a joint committee of the RGS and Royal Society.[33]

Markham won. Even though the most vocal Royal Society representative on the joint committee, Oxford entomologist Edward B. Poulton, took the battle to the pages of *Nature* and presented it as a conflict over whether scientists or navy officers controlled government research, Markham prevailed. He simply cared more about the expedition than anyone in the Royal Society and fought hardest for doing it his way. Contemptuous of opponents, Markham privately denounced Poulton as "totally ignorant on every subject that could possibly come before the Committee" and accused Royal Society members on the joint committee who called for scientists to set *Discovery*'s course as "trying to burn down *our* [the RGS's] house, to cook *their* eggs." Indeed, he dismissed the entire joint committee charged with overseeing *his* expedition as "too many cooks! actually 47: some asleep, some striking, some trying to spoil the broth."[34]

Before the Admiralty released navy officers for the *Discovery* expedition and while still working with the Royal Society to secure public funding for the venture, Markham had expressed little interest in the division of power between the ship's captain and the civilian scientists. Early planning documents suggest that a civilian science director would command the shore parties and perhaps also the scientific work at sea, subject to the captain's control over matters relating to the ship's safety, with the expedition likely depositing the science director, some scientists, and a few sailors on the Antarctic mainland for the winter and returning to retrieve them in the spring. As science director, the committee chose British Museum geologist J. W. Gregory, who had led a successful African expedition and served on one to the Arctic. Poulton depicted him as "the man who is, before all others, fitted to be the Scientific Leader of the National Antarctic Expedition," and even Markham initially conceded that "his scientific ability is undoubted."[35] Yet once navy officers became available and Markham got to pick his captain, he insisted on a traditional naval chain of command, with Gregory under Scott. Further, he insisted that the ship and entire crew winter in the Antarctic in the manner of earlier British Arctic expeditions. Markham now depicted Gregory as "an incompetent leader with no practice in dog driving" and dismissed his African expedition with native guides as "an independent journey, without [white] companions."[36] Of course, Scott had no experience driving dogs, leading an expedition, or conducting scientific research—he had not even commanded a ship or seen pack ice—but Markham saw in him the makings of a polar explorer.

Prolonged negotiations between the RGS and the Royal Society over the wording of the expedition's instructions became stalled over the issue of command. Hoping for a breakthrough, early in 1901 Poulton wrote a confidential letter to *Discovery*'s designated second officer, Albert Armitage, asking if he would assume command of the ship on the Royal Society's terms. "I have always contended that, as this is a

Scientific Expedition, it should be under Scientific leadership," Poulton wrote. "I need not say there will be a severe struggle over this." Refusing to mutiny, Armitage showed the letter to Scott. Markham used it to force a showdown with Poulton. "Scott was there," Armitage wrote: "He told me that the meeting was stormy, to say the least. Grave and reverend professors hurled epithets at Sir Clements, who presided, which astonished him. But the President of the R.G.S. held his ground. . . . He was a most determined, obstinate fighter of the real bulldog type."[37]

The sparring between the societies went on for months until some on the Royal Society side of the joint committee relented "in the interest of peace," as one of them told Poulton. Gregory promptly resigned. "Were I to accompany the expedition on those terms," he wrote, "there would be no guarantee to prevent the scientific work from being subordinated to naval adventure."[38] In some ways, it represented a battle between a nineteenth-century conception of scientific expeditions defended by the RGS and twentieth-century ones advanced by the Royal Society, with the older view prevailing one last time.

Gregory's assistant, British Museum botanist George Murray, became the expedition's science director but sailed only as far as Cape Town to train the remaining scientific staff. Just one member, twenty-five-year-old magnetic observer Louis Bernacchi from the *Southern Cross* expedition, had been to the Antarctic. He was a last-minute replacement for a more experienced physicist, William Shackleton, who by some accounts lost his post for siding with Gregory. The other scientists were Plymouth Museum marine biologist Thomas Hodgson, novice geologist H. T. Ferrar, and surgeons Reginald Koettlitz and Edward Wilson. Unlike naturalists from other expeditions, none of them became well known for their research. Yet they knew their place on this cruise. Markham said of Hodgson, "When he is sat on he takes it well." He added that Ferrar, who joined the staff straight out of college, "gets a good deal chaffed and sat on by the young lieutenants, which is already bearing fruit." While Markham seemed to respect Murray, the feeling was not mutual. In a letter to the RGS's Mill, who sailed on *Discovery* to Madeira instructing the staff in oceanography, Murray wrote about Markham: "That ridiculous person will make a mess of the expedition if left to himself."[39]

As the expedition unfolded, Scott effectively served as both captain and science director. "Associated with you, but under your command, there will be a civilian scientific staff, with a Director at their head," the captain's final instructions stated, but after Murray left the ship in Cape Town, the staff had no other leader than Scott.[40] Recognizing the importance of science for a successful mission, Scott took a keen avocational interest in it and tried to learn what he could from Murray about the research program. A quick study, he supported the scientific researchers and facilitated their work.

Although the expedition's instructions gave primacy to magnetic research and geographic discovery, they did not neglect John Murray's interests from *Challenger* days. "You will see that the meteorological observations are regularly taken every two hours," they stated. "Whenever it is possible, while at sea, deep-sea sounding should be taken with serial temperatures and samples of sea-water at various depths. . . . Dredging operations are to be carried out as frequently as possible."[41] In executing these orders, Scott clearly appreciated the difference between the duty imposed for meteorology and the discretion allowed for oceanography.

John Murray's vision of *Discovery*'s voyage as a second *Challenger* expedition never materialized. The ship was partly to blame. Although built for scientific exploration and loaded with gear for oceanographic research, it leaked badly, sailed poorly, and consumed excessive coal under steam. Falling behind in its southbound voyage, the expedition did not stop for sounding or dredging and skipped a planned stay in Melbourne, where it was supposed to coordinate research with a local observatory. Instead, *Discovery* sailed straight from Cape Town to New Zealand, where it was dry-docked twice for repairs, and departed for the ice two weeks behind schedule. "I should like to have the builder's people at sea for a few days," Scott wrote from New Zealand. "We have an account to settle for their basically scrapped work." Water leaking into the bilges rotted the nets used to collect marine specimens. "As this bade fair to continue," Hodgson noted, "all the nets were tarred in New Zealand as an attempt to preserve them: it was successful in this respect, but it depreciated their value for work in cold temperatures. They became so hard as to be difficult to manipulate, and also inflicted far more injury on the specimens than if they had been treated in another way."[42]

Unwilling to slow the ship for research until it reached the Ross Sea, Scott tried to improvise under sail. "Our original tow-nets were designed for use only when the ship was drifting," he wrote. "By increasing the length of the net and largely reducing its aperture, we found that we could use it whilst the *Discovery* was traveling through the water at her ordinary speed." This worked somewhat for plankton, Scott noted, but not for crustaceans: "Mr. Hodgson, our biologist, in whose department these were, reported that the delicate organisms were hopelessly destroyed, and came up 'all heads and tails.'" The original plan called for ocean research during the second summer, while a shore party worked on land, but *Discovery* remained icebound at its McMurdo moorings. In designing the ship, Markham had rejected the rounded hull of such modern, steam-powered polar vessels as *Fram* and *Gauss,* which could ride over ice, in favor of a traditional British V-shaped bow, which had to cut through it. "This decision contributed to the besetting of the *Discovery* for two winters," historian T. H. Baughman concluded.[43]

In published reports of the expedition's findings, Hodgson summed up the biological research at sea: "As a matter of fact very little was done." During *Discovery*'s voyage through the Southern Ocean, the expedition made even less effort to study deep-sea temperatures and currents than to collect biological specimens. Aboard *Gauss* for the concurrent German expedition, Erich von Drygalski was able to gather better oceanographic data and deep-sea specimens than *Discovery*'s researchers. By fitting his findings into those of other expeditions, he "confirmed" the theory of an undercurrent stabilizing the world's deep-sea basins with a slow, steady outflow of very cold bottom water from the Antarctic.[44]

A loner at heart and compulsive worker, Hodgson tried to compensate for the lack of biological collections on the high seas by ones taken around winter quarters, but here too he was beset by obstacles. A sheet of ice up to eight feet thick covered the sea during most of the expedition's twenty-four months at McMurdo Sound. "Of course the ice sheet reduced very considerably the area of operations," Hodgson wrote. "Open water and a boat would have enormously increased the collections, and though the *Discovery* was in winter quarters six weeks before the sea was effectively closed, that was a busy period, and it was only at intervals that a boat's crew could be obtained."[45]

Once stable, the sea ice provided somewhat of a research platform for Hodgson. "As soon as the Sound was frozen over the biological work was carried on through holes in the ice," he wrote. "My two principal holes were a mile and a half mile from Hut Point, in 125 fathoms."[46] He maintained more distant holes in deeper water. After sawing an opening, he would lower a baited trap or let a tow-net drift in the current for a day or two. By then, up to two feet of new ice would cap the hole, and he would have to cut it out to raise the trap or net. Ice crystals coated everything down to eight fathoms—deeper if the line remained submerged longer—spoiling any specimens they encased. "It should be borne in mind that a large portion of the work was done in the dark" during winter, Hodgson noted, in air temperatures that dipped below −40°F. "Lines cannot be coiled or wound on a winch; when frozen they snap only too readily. It was always necessary to 'walk away' with them when hauled, and let them lay out straight on the flow." He took advantage of occasional cracks in the ice to extend a line along the opening and drop the middle with a tow-net or dredge attached so that he could drag it through the water or across the bottom.

Once he raised his specimens, Hodgson sorted them by type into watertight bottles and tins. "In the winter of course everything froze at once and had to be thawed out on board ship," he explained. "In the summer the specimens were hardly so well off, for, although they did not freeze, the water was generally full of ice crystals which, with the jolting of the sledge as it travelled shipwards, cut the more delicate specimens to pieces."

Upon *Discovery*'s return, Scott distributed the marine collections among more than three dozen naturalists who described them in five volumes published by the British Museum. While commending Hodgson's diligence in adversity, many of the armchair researchers disparaged the narrow scope and low quality of the collections. Nearly everything came from one small area of McMurdo Sound. One biologist complained about the jellyfish that "some of the specimens were in such bad condition that without a good clue it would have been impossible to determine the genus."[47] Another wrote of trying to fit pieces together to make whole specimens. Compared to collections from *Challenger* or even *Gauss*, which was icebound in deep water, scientists identified many fewer new species from *Discovery*'s haul. They also did not detect the hoped-for overlap with the organisms of earlier geological periods or the Arctic. To fill out their surveys of Ross Sea fauna, naturalists turned to specimens from the *Southern*

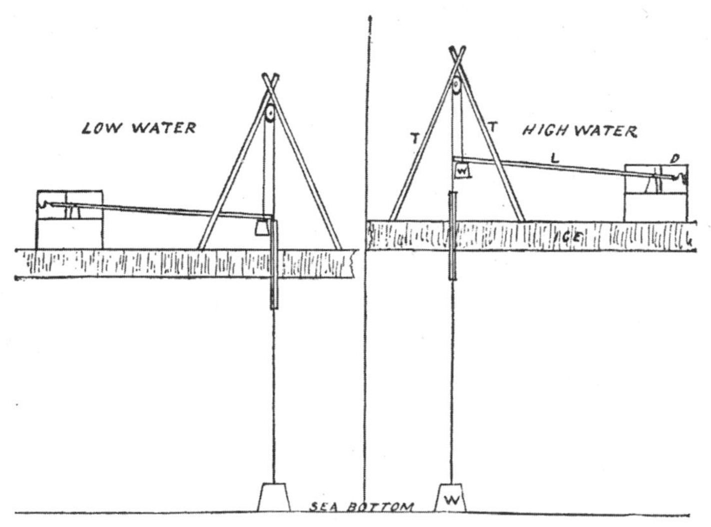

DIAGRAM SHOWING THE PRINCIPAL PARTS OF THE TIDE-GAUGE

W = the weights, the larger one as anchor, lying on the sea-bottom, the smaller one (the object of which is to keep the line taut at all times) on the free end of the line, below the lever; L = the long bamboo lever for reducing the scale; T = two legs of the bamboo tripod, supporting the pulley over which the wire passes; D = the drum on which the record is made. The recording part of the apparatus is more complicated than the diagram indicates. It will be described in detail in the scientific publications.

Diagram of a gauge for measuring the tide raising and lowering the sea ice covering McMurdo Sound during the Antarctic winter, from Ernest Shackleton's *Heart of the Antarctic* (Philadelphia, 1909).

Cross expedition—which must have vexed Markham, who dismissed that rival enterprise as a meaningless media stunt. Nevertheless, the expedition confirmed that McMurdo Sound—and by extension Ross Sea—was rich in animal life and deficient in plants.

Besides collecting marine specimens, the explorers also used ice holes to track ocean tides and temperatures during their stay at Hut Point. The tides, they found, rose and fell daily with seasonal irregularities. Water temperature in the sound hovered a degree so below

0°C, or just above freezing for salt water. While the expedition expended sustained effort on such matters, its oceanographic findings fell far below expectations. "The reason is obvious," Scott explained later, "as the greater part of our time was spent either locked in the ice or cruising in shallow seas; yet, as I look at the vast amount of this work which remains to be done in this area, I cannot but regret that we were unable to effect more."[48]

In the shuffle following Gregory's resignation as science director, Scott assigned meteorology to First Lieutenant Charles Royds, who had only a naval officer's training in the field. By doing so, Scott did not intend to demote that research. Indeed, he later wrote, "To obtain a complete record of meteorological observations was one of the most important scientific objects of the expedition," adding that "the prevailing direction of the winds has naturally an important bearing on the general circulation of the atmosphere in the Southern Regions."[49] He simply believed that a good navy officer could keep weather records and wanted as few civilians on board as possible.

The expedition's instructions called for weather observations every two hours, which required someone to check instruments around the clock. At sea, an officer always stood watch, but at Hut Point this meant special duty. Some meteorological instruments remained on the ship, where everyone lived in relatively warm but smoky conditions; others were moved 120 yards away behind raised, protective screens on the sea ice, where they were subject to the Antarctic climate. Because many of them were manually read, someone had to visit them. Royds made the rounds every two hours from 10 a.m. to 10 p.m. Wilson pitched in at 8 a.m. each day. "Night" duty rotated among other officers and scientists. "On a fine night this was no great hardship," Scott noted, "but in stormy weather the task was not coveted by anyone."[50] The readings took about twenty minutes in good weather—much more in bad. Ernest Shackleton waxed poetic about the chore in the expedition's journal:

> To be aroused from slumber at the deadest of the night,
> To take an observation, gives us all a morbid blight;
> How in the name of all that's blank, can temperatures down here,
> Concern those scientific men at home from year to year?
> To us alone they matter, for it's cold enough, Alas!
> To freeze the tail and fingers off a monkey made of brass.[51]

Except during severe blizzards, the expedition kept up this routine for over two years.

Under Royds's supervision, officers and scientists also regularly noted the temperature, air pressure, and wind conditions at various sites around Hut Point and during sledge trips. "The way he took charge of the meteorology," Markham said, is "enough to stamp him as a first class worker."[52] Certainly his efforts produced a mass of raw data, but perhaps still smarting from losing their choice for expedition science director, the scientists at the Royal Society were dissatisfied.

The division of duties for analyzing expedition findings gave meteorology to the Royal Society. At the time, scientists disagreed over how Antarctic weather fit into global patterns. All agreed that westerly winds, low air pressure, and high rainfall predominated between latitudes 40° and 60° south—classic cyclonic conditions for the Southern Hemisphere. Some thought these conditions persisted farther south. Others proposed that while westerlies might continue in the upper atmosphere, the low-level pattern was reversed with easterly surface winds, high pressure, and low snowfall across Antarctica—classic anticyclonic conditions. These models carried different implications for interpreting and forecasting weather in the Southern Hemisphere. With the models generating such dissimilar predictions for surface conditions on the Antarctic mainland, scientists hoped that the *Discovery* expedition could settle the matter. When it didn't, they blamed the messenger.

"Care was taken to draw up special instructions for working the self-recording instruments in order to secure correct timing and accurate readings," the Royal Society's report on the expedition's meteorological findings began. "Unfortunately, in consequence of changes in the staff, these well-intended efforts failed altogether." Problems plagued Royds's use of the new equipment devised for the expedition, forcing the young officer to rely on familiar instruments. This approach could misfire, for instance when he used mercury thermometers (included only as auxiliary equipment) rather than low-temperature ones, to take the regular readings, even though mercury freezes at −38°F, which was often exceeded. Further, the Royal Society report complained, "It was never contemplated that the station barometers would, as a rule, be so exposed to fail because the attached thermometers were not graded below 0°F." Even Scott realized something was amiss when the snowfall gauge drifted over in blizzards. "Somewhere far beneath the present surface lies the snow-gauge—a fact that makes comment on the utility of that instrument unnecessary," he noted in his diary after the expedition's first severe winter storm.[53]

Royal Society analysts grumbled most about the expedition's wind and air pressure records, which they had hoped would resolve disputes about the Antarctic climate. Surface winds at winter quarters generally blew from the east or southeast, which fit the anticyclone model for Antarctic weather, but Scott saw this as a local phenomenon caused by nearby mountains. To support his view, Scott pointed to findings from an ambitious thirty-one-day sledge journey by Royds, Bernacchi, and four others southeast across the Ice Barrier, which they made for the express purpose of obtaining meterological and magnetic readings unaffected by land.

Royds's records from this journey reported winds from the southwest, which supported the cyclone model for Antarctica's climate.

The Ice Barrier, however, is a featureless plane. Royal Society analysts determined that if Royds recorded the raw compass readings for these winds without correcting for variations caused by the region's skewed magnetic field, his records actually showed the wind blowing from the east, as at Hut Point. When pressed on the matter, Royds equivocated before ultimately affirming that he had made the correction. Scott backed his man, but proponents of the anticyclone model expressed their doubts. "How are we to account for the existence of a steady south-westerly current within a few miles of the ship," the Royal Society report asked, while "at the ship, although quite open and unsheltered towards the south-west, easterly winds were experienced almost exclusively?"[54]

Air pressure readings from the sledge journey could have helped to resolve the dispute, but again Royds's methods fell short. He found lower pressure as he traveled south over the Ice Barrier, which fit the cyclone model, but he did not adjust for changes in altitude. The Ice Barrier looked "level" to him.[55] One Royal Society analyst calculated that if the Ice Barrier sloped only two feet per mile toward the sea ("which is probably well within the actual amount"), then the relative pressure, corrected for altitude, actually increased southward.[56] As recorded in the Royal Society report, Scott responded by observing, "The staff possessed no means of discriminating between the influence on the barometer of the varying conditions of atmospheric pressure, on the one hand, and the difference in altitude on the other." By saying so, he conceded that the findings were useless. "It is curious," the Royal Society report stated, "that endeavors to reach, by two separate crucial tests, a definite conclusion upon this interesting point, as to whether the easterly wind at winter quarters is a local wind or a true general wind implying a high pressure to the south, fail through very slight omissions in the observations or the records." Gregory would have done better, the report all but said.[57]

* * *

In 1907, Shackleton returned to McMurdo Sound on the *Nimrod* determined to best Scott's *Discovery* expedition in every respect—including science. In Edgeworth David, he had a science director and senior geologist with research credentials and expedition experience to match Gregory's. With Douglas Mawson and Raymond Priestley, he gained young researchers destined to establish lasting reputations in science. The expedition's second-in-command, naval reserve officer Jamison Adams, served also as its meteorologist. On the advice of *Challenger*'s John Murray and polar explorer William Speirs Bruce, Shackleton chose the hard-working, self-educated scientist James Murray for biology. An adventuresome twenty-year-old heir to a vast landed estate, Baronet Philip Brocklehurst, went along as assistant geologist in return for a £2,000 donation to the expedition. A product of the merchant marine, Shackleton cared little about navy discipline and welcomed civilians on board. "He had assembled a powerful scientific team almost despite himself," historian Beau Riffenburgh concluded.[58] Without the scientific pretensions that hobbled the big-budget, publicly funded *Discovery* expedition, Shackleton's low-cost, privately financed enterprise easily exceeded expectations.

In the *Nimrod* expedition, Murray filled the role of *Discovery*'s Hodgson and used many similar techniques. Assisted by Priestley, the expedition's junior geologist, he spent much of the winter trapping marine specimens through holes in the ice and dredging the bottom along tide cracks. "It rarely happened that we found the cracks open," Murray wrote. "Usually they were filled with new ice to a depth of 6 inches or a foot, and it was by hard labour with ice-picks and crow-bar that we got a sufficient length open to serve for dredging. . . . To avoid dredging too frequently over the same ground it was necessary to cut trenches in the ice alongside the ends of the rope and at right angles to the line joining the two ends. In these trenches the rope could be shifted a yard or so at each time of dredging." The dredge and trap brought up vast numbers of the same few species: bivalves, snails, corals, starfish, sea urchins, large worms, sponges, sea anemones, and five types of big-headed fishes. "The sea bottom here appears to be covered by a continuous carpet of living things," Murray observed.[59]

Murray's collection gave scientists some appreciation of the diversity and distribution of Antarctic marine life. "Though our location at Cape Royds was only twenty miles north of the *Discovery* winter quarters," he wrote, "the local conditions differ very considerably. The temperature appears to be usually ten degrees or more (Fahrenheit) higher than at Hut Point. Being close to the spot where McMurdo Sound opens into the Ross Sea, we had open water close by throughout the year." This kept Murray near shore. Most of his specimens came from either a shallow, sheltered bay with a black-mud bottom or a steep bank with strong currents. "The collections differed a good deal in their composition," he wrote of his haul from the two sites, but not so much as the disparity between what he found and what Hodgson had reported. "This amount of difference between collections made at stations only twenty miles apart is very considerable. It is more important that one of our most abundant shallow-water species (*Yoldia eightsi*) did not occur at all at Hut Point."[60]

In other oceanographic and meteorological activities, the *Nimrod* expedition attempted to copy or improve on its predecessor. It tow-netted for plankton in the Ross Sea, charted tides and currents in McMurdo Sound through the ice, recorded weather conditions every two hours, and reported its findings in the same sort of heavy, bound volumes that memorialized the *Discovery*'s voyage. In an implicit reference to Royds's missteps, Shackleton noted in his popular account of the expedition that *his* men used spirit thermometers to measure temperatures below the freezing point of mercury and devised a stove-pipe gauge to measure snowfall in blizzards. For an expedition known and remembered mostly for Shackleton's dramatic dash toward the South Pole, its members devoted an extraordinary amount of time and effort to scientific fact-gathering. Science made the overall enterprise respectable in Edwardian England and helped secure Shackleton's knighthood.

The expedition's ascent of Mount Erebus showed how science gave meaning to adventure. Rising nearly 12,500 feet above Ross Island, Erebus not only towered over the explorers in their winter quarters at Hut Point and Cape Royds, tempting them to climb it, but also served as a high-altitude weather vane. "The proximity of Mount Erebus was a great stroke of luck," Scott wrote, "as the smoke of that volcano gave us an indication of the direction of the upper air currents." His expedition's Erebus smoke observations—368 in total, with three out of four showing winds from the southwest quadrant—helped convince him that easterly surface winds were a local aberration. Yet both models of Antarctic weather allowed for westerly winds spiraling inward toward the pole at upper altitudes, and proponents of the low-level anticyclone model noted that the expedition's cloud observations suggested a gradual shift from easterly and southerly low- and mid-level winds to higher westerlies. "The drift of the Erebus smoke would represent the motion of the air at a level approximately midway between the two," a Royal Society analyst noted.[61] During the *Discovery* expedition, Scott considered but rejected an ascent of Erebus. Shackleton's men stole that glory in the name of science.

The seven-day climb occurred in March 1908, after the *Nimrod* expedition had settled into winter quarters but before the full force of winter hit. "The observations of temperature and wind currents at the summit of this great mountain would have an important bearing on the movements of the upper air, a meteorological problem as yet but imperfectly understood," Shackleton explained, "and apart from scientific considerations, the ascent of a mountain over 13,000 ft. [sic] in height, situated so far south, would be a matter of pleasurable excitement."[62] He chose David, Mawson, and Forbes Mackay to summit, with Adams, Brocklehurst, and Eric Marshall in support. In the end, all attempted to summit despite the support party's lack of knapsacks and climbing gear.

Although Cape Royds sits on the flanks of Mount Erebus only seventeen miles from the summit, the climb was complicated by a steeply rising grade, bitter cold, and inadequate preparations. "At one spot," Shackleton wrote, "the party had a hard struggle, mostly on their hands and knees, in their effort to drag the sledge up the surface of smooth blue ice." At others, deep sastrugi, or wind furrows in the crusted snow, made pulling uphill almost impossible. On day three, the climbers exchanged their 560-pound sledge for 40-pound knapsacks—with the support party improvising packs. "Some of us with our sleeping bags hanging down our backs," David wrote, "resemble the scorpion men of the Assyrian sculpture: others marched with their household goods done up in the form of huge sausages."[63] That night, gale-force southeast winds pinned them in collapsed tents for thirty-two hours at nine thousand feet, which was enough to convince them that surface-level blizzards could reach high into the atmosphere. Climbing in misfit ski shoes, on his twenty-first birthday Brocklehurst began developing a serious case of frostbite that later cost him a toe. He remained behind on the fifth day while the others pushed on to the summit. Near the top they entered a fairyland of whimsically shaped, hollow ice-mounds formed when hot steam rising from volcanic vents hit the frigid Antarctic air. No one had seen such a sight; it took David to explain it.

The party reached the volcano's rim on the morning of the sixth day. "We stood on the verge of a vast abyss," David and Adams wrote. "Mawson's angular measurement made the depth 900 ft. and the greatest width about half a mile." With their injured companion waiting at the last campsite, the climbers quickly headed back down. "Finding an almost endless succession of snow slopes below us, we let ourselves go again and again in a series of wild rushes," David recalled.[64] Tossing their packs ahead and using ice-axes like rudders, they slid over halfway down the mountain by 10 p.m., to where they had left the sledge. Marshall described the plunge: "Pushing bag, glissading, following up, recovering it, dragging, shoving, soaked through." All six climbers stumbled into winter quarters before noon on the next day. "Bruised all over," Marshall wrote, "nearly dead."[65] They were greeted with champagne and Quaker Oats. Along with reaching the South Magnetic Pole and Shackleton's Farthest South, their ascent became one of the expedition's three best-known feats. "Fierce was the fight to gain that bright height," Shackleton wrote in a poem commemorating the event.[66]

The findings of the *Discovery* and *Nimrod* expeditions yielded a consensus understanding of Antarctic weather. Reports of persistent easterly and southeasterly winds at Hut Point and Cape Royds bolstered the view of a low-level anticyclone characterized by relatively high pressure and generally low precipitation. Between about five thousand and fifteen thousand feet, David and Adams reported that the wind typically shifted more to the south. "This is the return current of air blowing back from the South Pole towards the Equator," they asserted.[67] Above fifteen thousand feet, they detected the northerly winds that fed the polar high pressure and completed the Southern Hemisphere's weather cycle.

Summarizing these findings in a later report, David and Priestley depicted the region's general weather pattern as a "great Antarctic high-level cyclone overlaying the permanent anti-cyclone." This combination represented the predictable pattern if the area was covered by

Diagram of the plume from an eruption of Mount Erebus suggesting the impact of upper-level wind currents, from Ernest Shackleton's *Heart of the Antarctic* (Philadelphia, 1909).

a low ice sheet rather than a high plateau, they noted: an upper-level cyclone circulating air south and a lower-level anticyclone cycling it back north. The actual mix of local surface winds "is much complicated by the rock and ice dome of the Antarctic taking the place of so much of what otherwise would have been the lower part of this anticyclone." The pole, they wrote, "appears to be the eye of the great anticyclone." Turning to the upper levels, they added, "Some have doubted the existence of this cyclone, but from the fact that whenever there was an extra powerful eruption of Erebus, so that its steam cloud was carried to an altitude of above 20,000 feet, we invariably noticed that it was caught by a powerful W.N.W. or N.W. current, we are inclined to believe that this huge permanent cyclone really exists." All this, they wrote, combine with the steep grade in slope and temperature from the Polar Plateau to the sea and the vast extent of the surrounding ocean "to make the Antarctic the home of winds of a violence and persistence without precedent in any other part of the world."[68]

For *Nimrod's* naturalists, Mount Erebus provided clues to begin resolving the puzzles posed by the continent's extraordinary weather. "Not only had we the great cone of Erebus to serve as a graduated scale against which we could read off the heights of the various air currents," David and Adams wrote, "but we also had the magnificent steam column in the mountain itself, which by its swaying from side to side indicated exactly the direction of movement in the higher atmosphere." When violent eruptions sent the steam over twenty thousand feet, "it penetrated far above the level of a current of air from the pole northwards, so that its summit came well within the sweep of the higher wind blowing in a southerly direction," the two researchers noted.[69] Firsthand experience with sastruga patterns and wind conditions on Erebus reinforced their observations of cloud and steam movements. Much work remained, but the Antarctic sea and sky finally began to fit into global patterns. Alluding to the mountain where Moses first saw the Promised Land, David and Adams concluded, "It would be hard to overestimate the scientific importance of knowledge of the meteorological conditions obtained at Erebus. Erebus is the Pisgah of the meteorologist."[70]

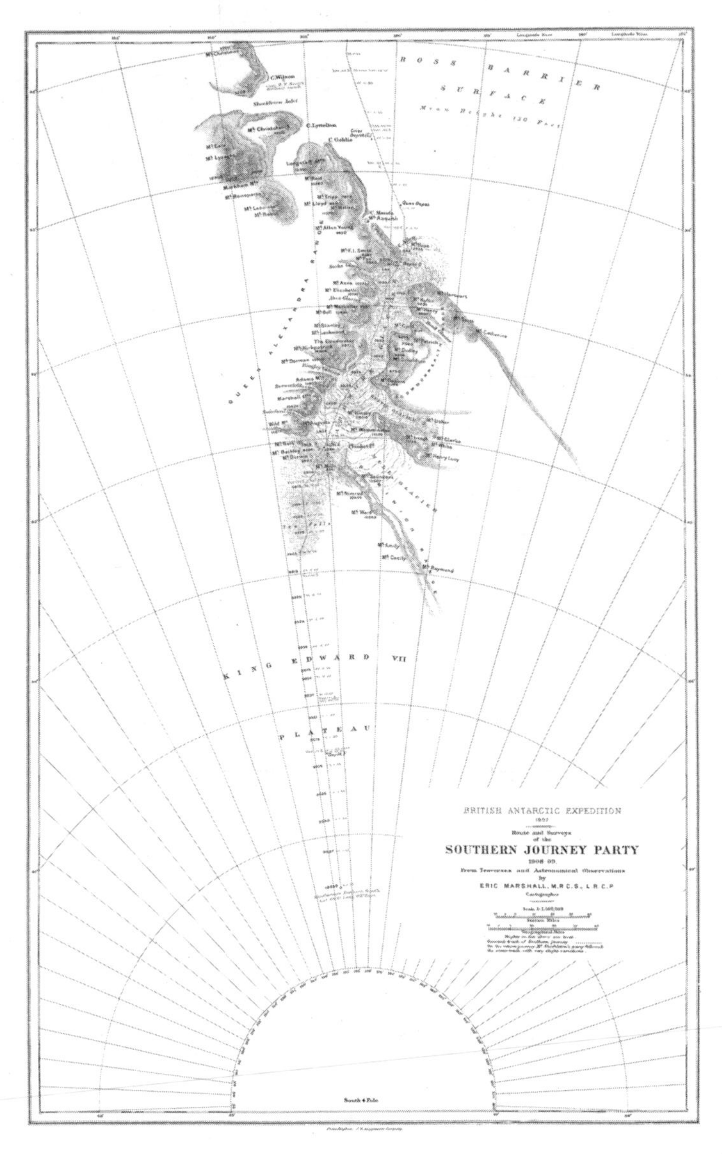

Southern Journey of Ernest Shackleton, Jameson Adams, Eric Marshall, and Frank Wild (1908–9), from Shackleton's *Heart of the Atlantic* (Philadelphia, 1909).

CHAPTER 5

Taking the Measure of Men

"THIS MUST BE OUR MOST NORTHERN CAMP. WITH five out of our little force totally prostrate, and four others exhibiting decided symptoms of the same complaint, it would be folly to persist."[1]

Albert H. Markham wrote these lines in May 1876 at latitude 83° 20′ north. Leader of the Northern Sledge Party for the first British naval expedition dispatched since 1827 with the stated objective of reaching the North Pole, Markham with three man-hauled sledges and a seventeen-person team had beaten the forty-nine-year-old Farthest North mark of William Edward Parry by 40 miles—yet fallen 460 miles short of his goal. "With this we must be content, having failed so lamentably in attaining a high northern latitude. It is a bitter end to our aspirations."[2]

Scurvy had stricken Markham's men as well as many other members of the 1875–76 expedition led by George Nares, forcing another humiliating British retreat from the far north. Few expected this well-provisioned venture to fail, but by the mid-nineteenth century heroic failure had become the refrain for British Arctic exploration. Attempting to follow in the wake of such legendary English navigators as Martin Frobisher, Henry Hudson, and William Baffin, who had opened the American Arctic to European discovery in their search for the Northwest Passage during the early 1600s, the Royal Navy was instead plagued by a string of setbacks. Those failures and the concerns that they raised about the flagging British spirit became a critical backdrop for later Antarctic endeavors.

The problems began with Parry but grew much worse. After initial success sailing two warships more than halfway across the Northwest Passage in 1820, Parry's next three expeditions failed to achieve their stated goals, including a spectacular reversal in 1827, when the floating pack ice north of Spitsbergen, which he hoped to cross on foot to get to the North Pole, carried his party south faster than it could march north. John Ross also fell short. His 1818 voyage merely retraced Baffin's two-hundred-year-old route, and the side-wheeled steamship used on his 1829–33 expedition was trapped in the ice for four years before he finally escaped with his surviving men on longboats left from one of Parry's earlier shipwrecks. Then came the traumatic disappearance of HMS *Erebus* and *Terror* under John Franklin's command in 1845, which led to a multiexpedition, ten-year search effort that mapped much of the Arctic Archipelago and logged about forty thousand miles of sledge travel before an effort led by Leopold McClintock finally found a trail of human remains and written confirmation of Franklin's death.

The Franklin searches may have fascinated the public and perfected the grueling art of man-hauled sledging, but it soured the Admiralty on Arctic exploration. Following McClintock's return in 1859, the navy did not dispatch another Arctic expedition until 1875, by which time the initiative had passed to other nations. Britain's lethargy did not sit well with the increasingly influential navy officers and former officers who romanticized their youthful service in the Franklin searches, particularly McClintock, Sherard Osborn, Erasmus Ommanney, George Nares, and Clements Markham. In 1865, along with Royal Society President Edward Sabine, who had served in the Arctic under Ross and Parry, they began plotting the Royal Navy's return to the Arctic.

By then, though no one had yet navigated the Northwest Passage, geographers generally agreed that there was little commercial or scientific purpose to doing so. Explorers had surveyed virtually the entire route and concluded that merchant ships could not use it. While conceding that it too would have no commercial and uncertain scientific payoff, the Arctic lobby made planting the Union Jack at the North Pole its new goal. "It is the only thing in the world that is left undone, whereby a notable mind might be made famous and fortunate," Markham declared at an RGS meeting in January 1865.[3] He would later make similar claims for Antarctic exploration.

Later that year, Osborn launched the campaign with a widely publicized address to the RGS. His "brother Arctic explorers do not belong to the new school of 'rest and be thankful' men," he snarled, and "are no more prepared to turn their backs upon the Arctic regions because Franklin died off King William's Land, than you would wish them to do so to an enemy's fleet, because Nelson fell at Trafalgar." Critics, he said, exaggerated the risks: "More sailors have been thrown to the sharks from the diseases incident to service in China and the coast of Africa, within the last four years, than ever fell in thirty years of Arctic service." Parry's Spitsbergen route might have failed, Osborn explained, but the Smith Sound passage between Greenland and Ellesmere Island promised easy access to the pole. Here, whalers had found a sea lane open in summer at least to Cape Parry at about latitude 82° north, with prospect of more land farther north. "The distance of Cape Parry to the Pole and back," he noted, "is just 968 miles; a distance which has been repeatedly exceeded by our Arctic sledge and boat parties since the year 1850."[4]

The plan Osborn proposed was the one that the 1875–76 expedition later attempted to follow. Two navy steamships provisioned for two winters would sail as far north as open water permitted in summer and lay provisions for the following year. Both would winter in the Arctic, one as far north as possible and the other somewhat south to serve as an escape vessel should the first become icebound. Sledge parties would set out in the spring from the northern ship, with one heading toward the pole and others surveying the coast. Osborn minimized the danger: McClintock had sledged 1,330 miles in 1859, he stated, and "agrees with me in thinking it is quite possible . . . to take a sledge from Cape Parry to the Pole and back." While suggesting that the effort might yield scientific benefits, Osborn stressed the psychic ones. "The Navy needs some action to wake it up from the sloth of routine, and save it from the canker of prolonged peace," he said. "Polar exploration is more wholesome for it . . . than any more petty wars with savages."[5]

Following Osborn's address, a dozen RGS members spoke in favor of the project, including Murchison, Markham, Sabine, and a physician from the Franklin searches who hailed "the salubrity of the Arctic regions" and discounted any risk of disease or disability other than frostbite. "To reach the Pole is the greatest geographical achievement which can be attempted," Sabine declared, "and I own I should grieve if it should be first accomplished by any other than an Englishman." Invoking her husband's work and the long search for his lost expedition, Franklin's widow depicted the pole as "the birthright and just inheritance of those who have gone through fifteen years of toil and risk in the Arctic seas."[6] Among those present, only Henry Richards expressed reservations about Osborn's scheme. Clinging to the notion of an open summertime Arctic Sea beyond the pack ice, as theorized by German geographer Augustus Petermann, Richards favored trying to sail to the pole by way of Spitsbergen.

Richards served as the Admiralty's hydrographer from 1864 to 1874. While he held that post, no British expedition attempted the Smith Sound route despite persistent prodding by Osborn, Markham, and the RGS. When a friendly Conservative government took over Westminster in 1874, however, ten years of work by the Arctic lobby paid off handsomely. Citing "its chances of success as well as the importance of encouraging that spirit of maritime enterprise which has ever distinguished the English people," Prime Minister Benjamin Disraeli directed the Admiralty "to lose no time in organizing a suitable expedition" to explore "the region of the North Pole." Nares was recalled from the *Challenger* expedition to command the mission, and the Admiralty had two steamships ready by May 1875, only six months after receiving Disraeli's orders. The hypothesis of an open Arctic Sea having by this time been discredited, even Richards now supported the Smith Sound effort. He joined Osborn and McClintock in drafting the navy's instructions for it. "The scope and primary object of the Expedition should be to attain the highest northern latitudes," they stated, "and, if possible, to reach the North Pole; and from winter quarters to explore the adjacent coasts within the reach of travelling parties."[7]

The British people began speaking of the pole as if it were already obtained. They admired it, one newspaper reported, "as gaily as if it were the maypole of a village festival. The Pole, and nothing but the Pole, was the talk of the day." Speaking of the polar conquest to the BAAS in 1874, army topographer C. W. Wilson proclaimed, "It is to be done, and England ought to do it." Who could deny the British Empire, already girdling the globe with colonies, its polar crown? Over the caption "WAITING TO BE WON," the British weekly *Punch* in 1875 printed a full-page sketch of a virginal ice-princess enthroned at the North Pole with guardian polar bears. This alluring "Witch-Maiden," a gender-laden accompanying poem noted, "draws manly hearts with strange desire to lift her icy veil." Having described reaching the pole as "a certainty, so far as human calculation can make it so," Clements Markham accompanied the expedition as far as Greenland, with his cousin Albert tapped to lead the Northern Sledge Party. "Never before has a Polar Expedition been so perfectly equipped, provisioned, and provided for against all conceivable perils," the *Times* of London commented. "All has been done that skill, money, and foresight can accomplish to make the Expedition under Captain Nares successful."[8]

The expedition left Portsmouth with great fanfare on Saturday, May 29, 1875: the queen's birthday. "There are few towns in the country which can so suddenly assume a festive appearance as Portsmouth," the *Times* reported. "On Saturday the transformation was complete." Bunting hung from houses, flags flew from every pole, and banners bedecked ships in the harbor. The army garrison paraded in the commons, and the lords of the Admiralty arrived en masse from London along with thousands of well-wishers. At a banquet for departing officers, the Lord Mayor expressed his confidence that "all that individual energy, pluck, and endurance can do will be done." The queen wrote to Nares of her "trust that you may safely accomplish the important duty you have so bravely undertaken." Ships, docks, and shore were packed with cheering throngs as the expedition's vessels sailed out to sea. Yachts followed in their wake. "The multitudes who assembled at Portsmouth to witness the sailing of Captain Nares and his brave companions," the *Times* observed, "indicate pretty clearly that a true chord has been struck, and that the sympathies of all, from the Queen downwards, go with them. . . . Those who have witnessed the race-course at Epsom on the Derby Day assert that it falls far short of the scene of Saturday."[9] Britain, everyone bet, would win this race.

The expedition's failure shocked the nation. Bad luck, poor planning, and botched execution contributed to the outcome, with any one of them probably sufficient to cause it. Bad luck came in the form of not finding land farther north than that already known at roughly latitude 82°, where the advance boat wintered and from which the polar party departed. "The absence of any visible land to the northward was extremely discouraging," Nares wrote. Without the ability to cache supplies on land or follow level ice along a coast, he knew that Markham's sledge party could not reach the pole but felt that it must try for high latitude. The party might have gone more than sixty net miles north if they had not traveled as a single party without support from the northernmost point of land. Because they could not lay advance depots on ice floes, two officers and fifteen men—the equivalent of two sledge-crews—pulled three sledges overloaded with supplies and two small boats for a drag weight of 404 pounds per person. "This," Markham conceded, "would necessarily render the progress slow and tedious."[10] The botched execution later chastised by the press, Parliament, and the Admiralty consisted of not dispensing sufficient amounts of lime juice or other antiscorbutic. The results were horrific.

The party set out in early April, before the spring thaw. "The cold then was so intense as to deprive the majority of the party of sleep," Markham wrote. At night, the temperature dropped to −45°F outside and −25°F inside the tent. Sleeping bags froze "so hard as to resemble sheet-iron" and became "wet blankets" when thawed by body heat. The lengthening days, however, were much worse than the nights. The temperature never rose above freezing until mid-June. "For the first three or four weeks after leaving the ship our bacon was frozen so hard as to be almost uneatable," Markham wrote. "As a rule, we were assailed by an intolerable thirst, which we were unable to assuage for two reasons: first, that we could not afford sufficient fuel to condense extra water; and secondly, it was quite impossible to prevent the water in the bottles from being converted into ice."[11]

Man-hauling the sledges proved all but impossible. The sea ice consisted of old, uneven floes up to a mile in diameter repeatedly thawed, rammed together, and recongealed, leaving hummocks of upended ice up to fifty feet high and a quarter-mile wide. "Around these hummocks were deep snow drifts, through which we frequently floundered up to our waists," Markham noted. The party often needed to cut a path for sledges through the hummocks with picks and shovels. Even on the floes, deep snow forced the party to divide and relay the load, which resulted in a one-mile net gain for five, seven, or even nine miles pulled. "It is a succession of standing pulls," Markham wrote of one day's work. "One, two, three, haul! and very little result. Distance marched nine miles; made good one-and-a-half." Some snowdrifts swallowed the sledges whole. The hard crust forced the men to lift their feet fully out of the snow for each step. "On several occasions," Markham reported, "the men found it not only easier, but they could make better progress whilst dragging the sledges, by crawling on their hands and knees, than by dragging in the more orthodox manner." In the end, he calculated that his men had marched 600 miles for the 130-odd miles covered, with little to distract them from their backbreaking work. "Everything of the same uniform colour," he complained in his diary, "nothing but one sombrous, uneven, and irregular sea of snow and ice."[12]

Scurvy made the journey a living hell for some and a death march for others. The disease prostrated its first victim on Easter, less than two weeks into the ten-week ordeal. "We unanimously came to the conclusion that it was the most wretched and miserable Easter Sunday that any one of us had ever passed," Markham noted in his diary. From then on, some sick men were borne on sledges, increasing the drag weight and diminishing the pulling power. Markham did not recognize the symptoms at first, so completely had scurvy been eradicated from the navy, but slowly the cause became clear when the stricken men did not improve and the condition spread. "The interiors of our tents in the evening have more the appearance of hospitals than the habitations of strong working-men," he wrote on May 8. Still the party pulled on without relief. The sledges only carried two small bottles of lime juice each, all of which froze solid. Markham finally gave up on May 12, when he calculated the distance to the pole at just under four hundred geographical miles. No one had ever attained such a high latitude north or south. "On this being announced three cheers were given," he wrote; flags were raised, the national anthem sung, and a magnum of Scotch whiskey passed around.[13]

The return march became a race against death that one man lost. At least the party could pass through cuts in hummocks made on the outbound trek, but with spring, snow became slush, and sledges sometimes broke through the ice with sick men aboard. Blizzards brought wet snow that soaked everyone to the skin. "The snow-drifts are far deeper and more frequent now," Markham noted in his diary about retracing the party's old route, "pools of water were forming between the snow-drifts, and a large quantity of sludge was encountered."[14] Finally on June 7, with fewer than half the men still in harness, the party's sole healthy member sprinted ahead for help. A sailor died the next day. By the time relief arrived, only six men had strength enough to drag sledges, which they relayed with two men borne on each and the rest stumbling along on foot.

Death stalked the expedition's other sledge parties too, and scurvy spread to those who remained behind on the ships. Four died, and half the crew became disabled. The Royal Navy had not seen such a serious outbreak of the disease since lime juice entered the prescribed regimen a century earlier. Some attributed the outbreak to cold, darkness, or close quarters, but none of these explanations made sense even then. The lack of vitamin C, which fresh meat, vegetables, and fruit can supply, caused it. At the time, however, no one knew about vitamins. Physicians had simply stumbled on the link between scurvy and diet, but not everyone believed it. On this occasion, expedition officers had stopped serving lime juice when the ships entered winter quarters—as if they thought scurvy afflicted sailors only when at sea—though by this time new methods of bottling lime juice had diminished its effectiveness as an antiscorbutic anyway. For Nares, the disease became a compelling reason to abandon the mission. His expedition carried enough supplies for two years in the ice, but once the spring sledge journeys ended, they steamed home rather than endure a second winter in the Arctic. "Pole impracticable," Nares wired from his first port of call, and so it became for Britain.[15]

Clements Markham tried to put the best face on his cousin's efforts by hailing the party's Farthest North and claiming that no one cared about the pole. But they did. "People talked so much about the Pole that they came to think themselves quite injured in being unable to reach it," the *Times* wrote. The expedition, another London newspaper sneered, "went out like a rocket and came back like a stick." "I do not claim for the Expedition any brilliant success," Richards ultimately conceded. "It is probable that no one engaged in it does make this claim."[16]

Unwilling to risk another failure and uncertain of public support, Markham and other RGS leaders began looking south for polar glory. In his introduction to Nares's book about the expedition, Richards questioned whether British explorers would ever again attempt a sledge journey to the North Pole. "Geography has little to gained by it, science perhaps less," he wrote.[17] Antarctica, with solid land under foot, promised better prospects for sledging and science.

The expedition's failure became an occasion for national soul searching. Blasting the "careless bungling" that marked the venture, the *Saturday Review* observed, "There seems to be very few things the Admiralty does not more or less muddle." The *Times* agreed: "There have been many instances of Naval Mismanagement lately, but they would all be eclipsed by such inexcusable neglect as the omission to provide Arctic crews with an effective anti-scorbutic." Albert Markham vigorously denied that disease alone prevented his party from reaching the pole. Even a healthy sledge party would not have exceeded his mark "by many miles, certainly not by a degree," he asserted, and he joined his captain in asserting "the utter impracticability of reaching the North Pole."[18]

Such comments elicited protests from those who saw the failure as a sign of diminished national character or the decline of empire. "We ought not to have been told . . . that to reach the North Pole is 'impracticable,'" the *Gentleman's Magazine* opined. "The explorers did all that it was possible for them to do . . . but it does not follow that we shall never reach the Pole." The *Spectator* added, "England should not, after her magnificent efforts and in the face of the whole world, publicly acknowledge herself defeated by the dangers and difficulties of a march one-fifth of which has already been accomplished." Seeing the matter as an international competition, the *Saturday Review* counseled, "For the mere failure to reach the North Pole we can easily console ourselves . . . so long as nobody else can succeed where we have failed." Nares declared, "It is true that we failed to bring home the North Pole as a national present to the world, but those that regret that circumstance may be consoled with the knowledge that failure implants more deeply in all breasts the desire to excel."[19] For those concerned about national survival in a Darwinian world, the expedition served as a clarion call to action. Britain must push on or be passed by, they argued. Antarctica offered a proving ground for national fitness.

Even as Nares beat a hasty retreat from the Arctic, the British people faced a much graver threat to their collective self-confidence. By 1875, the German states had unified into a continental empire that threatened British political, military, and economic dominance in Europe and potentially the world. At the time, Britain had assembled the largest empire and strongest navy in history. The empire produced over half of the world's coal, iron, steel, cotton cloth, chemicals, and many other strategic goods. Its overseas trade exceeded that of the next three leading commercial nations combined. Money followed commerce, and British investors and banks financed and managed the global economy. Nevertheless, during the final quarter of the nineteenth century, a growing chorus of domestic Jeremiahs warned of imperial decline. Britain had become decadent, they feared. Wealth and ease had sapped the vigor of its people. Germany, the United States, and even Russia threatened to supplant Britain, much as it had supplanted Spain and France.

Politicians and pundits marshaled evidence that expressed and fed these fears. United Germany's population exceeded that of Britain by 25 percent in 1875 and grew faster due to the United Kingdom's lagging birthrate. By 1900, Germany had surpassed Britain in the production of steel, glass, synthetic dyes, electrical equipment, and many precision manufactured goods. It was rapidly catching up in international trade. Rather than boost national confidence, the British occupation of Egypt in 1882 to protect the Suez Canal, which consolidated world opinion against the United Kingdom's imperial ambitions, deepened domestic feelings of isolation and vulnerability. Pulp fiction and the penny press warned of an invasion of the British Isles by German or French forces—or both. Already possessing a much larger army than Britain, Germany launched a program to double the size of its navy in 1898 and again in 1900. British public opinion, long divided over colonial policy, coalesced around the idea of empire as the island nation's best hope to remain ahead of geographically larger powers like Germany, Austria-Hungary, Russia, and the United States. By the fin de siècle, many nationalists in the RGS and other establishment organizations viewed a revived national spirit and revitalized navy as essential to defending the homeland, maintaining overseas territories, and retaining Britain's dominance in sea trade.

Throughout the Victorian era, polar exploration had been portrayed as a means to train and discipline naval officers and sailors during peacetime. For example, while acknowledging in advance that "the utility of [the 1875–76 Arctic] Expedition was not very evident," RGS President Henry Rawlinson noted that "if the undertaking only served to stimulate the courage, daring, and boldness, and keep up the chivalry that has always distinguished the British navy, that would be quite sufficient to most Englishmen." After it returned without reaching the North Pole, Markham defined the expedition's principal saving grace as "the creation of a young generation of experienced Arctic officers."[20] As concerns about Britain and its navy grew, leaders of the polar lobby within the RGS and retired officers' corps sought to rekindle interest in Antarctic exploration. A successful expedition could revive the nation's spirit and toughen young naval officers, they argued, even if it did not result in the acquisition of inhabitable territory.

No one did more to raise the alarm about Britain's survival in a Darwinian world than the RGS's Francis Galton. An eccentric polymath, Galton explored southwest Africa during the 1850s and proposed the theory of Southern Hemispheric anticyclones that provided a theoretical framework for later Antarctic climate research. When his cousin Charles Darwin published *On the Origin of Species* in 1859, Galton instantly converted to evolutionism and adopted a deterministic view of heredity. From the mid-1860s until his death in 1911, he championed the cause of selective breeding for human improvement, for which he coined the term "eugenics" from the Greek for "well born."

Galton's embrace of eugenics stemmed from his anxiety over the decline of what he called the British race. The spread of state and charitable social services allowed the lower orders to survive and propagate, he reasoned, while educated elites opted to have ever fewer children. Prosperity had dulled the cutting edge of human natural selection in Britain. "So the race gradually deteriorates," Galton warned in his 1869 book *Hereditary Genius,* "until the time comes when the whole political and social fabric caves in." To address the problem, along with reproductively isolating the unfit, he proposed indentifying vigorous, able youth, giving them "a first-class education and entrance into professional life," and encouraging them to mate.[21]

Eugenics gradually gained ground in British academic and professional circles along with a conviction that Britain was insufficiently "efficient" to keep ahead of Germany in the international struggle for existence. During forty-plus years on the RGS Council and two terms presiding over the BAAS's Geography Section, Galton promoted his ideas among geographers and explorers. Markham carried some of them into his promotion of Antarctic exploration. Where he once favored old Arctic hands to lead polar expeditions, for example, he became the champion of young officers chosen for their physical and mental adroitness regardless of their experience on ice. "The fatal mistake, in selecting Commanders for former polar expeditions, has been to seek for experience instead of youth," he wrote, with Nares and Franklin apparently in mind. "Old men should supply information and the results of experience, and should stay at home, making way for the younger and therefore more efficient leaders. New ideas, novel situations meet with cordial welcome when young men are at the helm."[22] Like Galton, Markham now placed his faith in such men.

Scott filled Markham's requirements for leadership. Only thirty-three years old when *Discovery* sailed and without experience commanding a ship or sailing in ice, he impressed Markham as an ideal captain based on a plucky performance in a cutter race fourteen years earlier. "I believe him to be the best man for so great a trust, either in the navy or out of it," Markham asserted. This choice prevailed over stiff opposition within the organizing committee. "All our Arctic officers are now old men and, as youth is essential, one without actual Polar experience has had to be selected," the *Times* wrote in reporting Scott's appointment. "But that is a small matter; any competent young officer will soon make himself familiar with what has been done and what remains to do." As for the chosen officer, the *Times* added, "There can be little doubt that Lieutenant Scott will make the most of the splendid opportunity afforded him."[23]

Discovery's other officers fit a similar pattern. All of them volunteered for the mission. Except for Albert Armitage, all were under thirty and without polar experience. Despite the Admiralty's reluctance to release its officers for the venture, all but Armitage and Shackleton came from the navy. When listing their attributes, Markham used terms like "charming," "zealous," "well bred," and "exceedingly good tempered." For each, he designed a heraldic sledging flag. As obsessed with genealogy as Galton, in his private notes

Markham worked out that four of the ship's officers (including Scott) were related to him by blood or marriage. "I decided that he should be one of the Antarctic heroes," Markham wrote of twenty-five-year-old Charles Royds, who Markham determined was a first cousin once removed of the wife of a cousin whose wife was one of his own second cousins. He found it worth recording that Shackleton's great-grandfather "was the instructor of Edmund Burke." In his mind, such born leaders could guide Britain to polar glory. "Sherard Osborn and I were agreed upon an Antarctic Expedition," Markham wrote in a telltale passage. "Its main object would be the encouragement of maritime enterprise, and to afford opportunities for young naval officers to acquire valuable experiences and to perform deeds of derring doe."[24]

As *Discovery* prepared to sail, Britain's long-simmering concern about national competitiveness boiled over into a full-blown crisis of doubt about the empire's survival. In October 1899, only three months after the government authorized funds and navy personnel for the National Antarctic Expedition, Britain went to war with two small South African republics. The Transvaal and Orange Free State had scarcely one hundred thousand white Boers, mostly of Dutch descent, ruling over a much larger number of native people and a growing population of disenfranchised British settlers. Having been pushed out of the Cape Colony by the British during the 1830s, the Boers had entrenched themselves in the South African interior and were determined to fight further incursions. They succeeded during the First Boer War in 1880–81, when Britain opted not to engage in protracted warfare for a region of little apparent value; but the subsequent discovery of gold in the Transvaal brought more British settlers and demands that threatened Boer sovereignty. When the Boers responded with a declaration of war and preemptive raids into neighboring British colonies, the London press boasted that British forces would quickly subdue the two Lilliputian states, which had no regular defense force.

Instead, the vaunted British army faltered before a volunteer militia of farmers that never exceeded forty thousand men.

Despondency replaced overconfidence as British forces suffered a string of defeats in trying to repulse the Boer offensive. After a chaotic battle on October 30, 1899, more than a thousand British regulars gave up to a Boer force less than half that size at Nicholson's Nek—the largest surrender of British forces since the Napoleonic Wars. Mournful Monday, as it became known, gave way to Black Week in December 1899, when tactical errors and operational blunders led to the British losing nearly three thousand men killed, wounded, captured, or missing in three battles, versus only a few hundred Boer casualties. "The war news is terrible. Never have we been so low," the bishop of London wrote that week. Contrasting the response to these defeats with celebrations marking the sixtieth year of Queen Victoria's reign, the *Times* commented, "The war has been the nation's Recessional after all the pomp and show of the year of Jubilee."[25] More losses followed before Britain reversed the situation by dispatching an overwhelming force of some 250,000 soldiers against a militia one-tenth that size. Still, the British suffered two more years of war and resorted to scorched-earth tactics and the internment of noncombatants before securing an uneasy peace. By this time, world opinion had turned to favor the beleaguered Boers, and many in Britain worried what might happen if Germany decided to attack while all available troops were deployed in South Africa.

The Boer War administered a shock to the British psyche that the *Discovery* expedition would address. During the early months of the war, articles in leading British journals carried such titles as "The Darkest Hour for England" and "Collapse of England." A scathing article titled "A Nation of Amateurs" lamented the "many signs of weakness . . . forced upon public attention by the humiliating experience of the South African War." The army came in for near-universal criticism: it had collapsed "in less than sixteen weeks," one pundit noted, "after the outbreak of hostilities against two States, whose united population is less than that of Birmingham or Leeds." Worse still, when it tried to recruit soldiers for the war, most of those volunteering proved unfit for service. Recruiters took fewer than one out of ten volunteers into the regular army from Birmingham, for example, and those accepted nationwide tended to be shorter and smaller than in the past. "An Empire such as ours requires as its first condition an Imperial race—a race vigorous and industrious and intrepid," the Earl of Rosebery declared in 1900. "The survival of the fittest is an absolute truth in the conditions of the modern world."[26]

At the dawn of the new century, Britain needed to restore its self-confidence. "The spirits of one and all, whatever their political party or their opinion on the rights or wrongs of the British actions in South Africa might be, [are] depressed in a manner probably never before experienced by those of our countrymen now living," Galton's protégé, Karl Pearson, observed in 1900. "Before England and the Empire lies a future of danger and uncertainty," a *National Review* essayist added that same year. In 1901, Scottish-American industrialist Andrew Carnegie wrote that he had never known "such despondency" in London. By the time that *Discovery* sailed later that year, an essay in *Nineteenth Century* had noted, "England has never felt more acutely than during the past eighteen months the want of great men." Scott and Shackleton, with their amalgam of old-fashioned courage and newfangled scientific purpose, walked into that void. Commenting on the expedition that first united the two future national heroes, an editorial in the *Times* of London observed, "The combination of the spirit of adventure with devotion to the cause of science which inspires the Polar explorer will always appeal strongly to the countrymen of Frobisher and Franklin."[27]

Ambition drew the two men together and drove them apart. Neither had a native interest in polar exploration. Both saw it as a means of escape: Scott from the limited chance for promotion that the peacetime navy offered junior officers without means or family connections; Shackleton, as a close friend later recalled, "from an existence which might eventually strangle his individuality" through routine service in the merchant marine.[28] With the international race for the North Pole heating up and interest burgeoning in Antarctica, polar exploration offered a path to glory that both men grasped for like the brass ring of a Victorian merry-go-round. With Markham as his mentor, Scott needed only to ask for command. While serving on a passenger ship ferrying troops to the Boer War, Shackleton had befriended the South Africa–bound son of the expedition's chief benefactor, Llewellyn Longstaff, and through him gained a place on *Discovery*'s roster. Scott believed in naval hierarchy, Shackleton in individual enterprise. Scott was reserved, conventional, and surprisingly insecure, Shackleton exuberant, restless, and brimming with self-confidence. Both led by example and acquired devoted followers. With only one pole, even Antarctica was not big enough for both of them.

Both men faced a steep learning curve with failure a constant threat. Their inexperience became quickly apparent when the *Discovery* expedition attempted some preliminary sledge journeys early in 1902, before the onset of its first Antarctic winter.

The opportunity to lead the first overnight sledge trip from *Discovery*'s winter quarters fell to Shackleton by virtue of a coin toss with a fellow officer. The expedition's twenty-three-year-old geologist, Hartley Ferrar, and twenty-nine-year-old assistant surgeon, Edward Wilson, accompanied him. None had polar experience. Their destination was a nearby island that promised a better view of the Ice Barrier to the south. In the clear Antarctic air, the island appeared so close that members of the expedition had attempted to walk around it one afternoon only to find that hours of walking seemingly brought it no closer. Nevertheless, Shackleton, Ferrar, and Wilson eschewed skis and dogs for man-hauling an awkwardly packed sledge topped with a wooden skiff for use if the sea ice gave way. None of them knew how to set up the tent or had practiced using the paraffin stove in the field. They set off before noon on February 19 with their sledge flags flying.

Shackleton, greatly misjudging the distance, expected to reach the island by early evening. After eight hours of pulling in the face of a rising gale, the island still looked far away. The men kept going with only one short break and virtually no food until Shackleton called a halt near midnight. "The pulling was fairly hard," Ferrar wrote. "At the time we were hardly making 6 inches per step." Wilson detected frostbite on both his companions and quickly felt it himself when they stopped pulling. They had sweated profusely in their windproof gabardine garments. "No perspiration gets out through this, but you feel dry because it freezes," Wilson noted. They struggled to erect the tent, light the stove, and don their cumbersome fur night suits, which Wilson described as "simply awful." When the wind died down a few hours later, they resumed their march but camped again after four hours out of sheer exhaustion. They did not reach the island until late on the second day. From its summit, they saw the Ice Barrier's flat plain stretching endlessly south and east, with Victoria Land's coastal mountains extending southward in the west. The way to the pole would lie across the Ice Barrier near those mountains. Scott darkly reported that on their return, "our travellers shook their heads over the bright prospect of a smooth highway" to the south.[29]

The explorers had learned more about themselves than about the terrain. "Although the temperature had not been severe, our travellers had nearly got into serious trouble by continuing their march in a snowstorm," Scott noted. "They found themselves so exhausted when they did stop to camp that they were repeatedly frost-bitten. They could only get their tent up with great difficulty, and then followed all sorts of troubles with the novel cooking apparatus."[30] Worse trouble lay ahead.

* * *

Ten days after Shackleton's group returned, four officers and eight men with two dog teams embarked on a fifty-mile journey around the icebound side of Ross Island to Cape Crozier, where they hoped to post notice of *Discovery*'s location for any relief ship. Scott planned to lead the party, but a skiing accident forced him to put Royds in charge. "The sledges when packed presented an appearance of which we should afterwards have been wholly ashamed, and much the same might be said of the clothing worn by the sledgers," Scott later wrote. "Not a single article of the outfit had been tested, and amid the general ignorance that prevailed the lack of system was painfully apparent in everything."[31] Royds harnessed men and dogs together to pull the sledges because he did not realize that dogs will not pull with humans. In fact, no one knew how to work the dogs. Two weeks earlier, Armitage and Louis Bernacchi had engaged in a sledging race to see whether whipping or coaxing dogs worked better. The result was pandemonium among the teams with the prize going to Bernacchi only because Armitage never got the whipped dogs to pull at all.

Packing too much of this and too little of that, Royds set off in what historian David Crane depicted as a "chaotic caravan of men and dogs." After traversing Ross Island's narrow southern ridge, the party proceeded along the Ice Barrier but soon bogged down in deep, soft snow. "The labour was excessive, and the dogs were of no assistance," an expedition report stated.[32] After struggling with the snow for three days, Royds decided that skis were needed to cross it. The party had packed only three pairs, however, so he skied on with two fellow officers and sent the others back.

With twenty-three-year-old Second Lieutenant Michael Barne leading eight inexperienced men, the return party had barely reached the ridge when a blizzard struck. Visibility dropped to inches. The party wisely set up camp among some rocks but unwisely abandoned it as the storm got worse. "We afterward weathered many a gale in our staunch little tents," Scott later noted, "but to this party the experience was new; they expected each gust that swept down on them would bear the tents bodily away, and meanwhile the chill air crept through their leather boots and ill-considered clothing." Some had frostbite; two wore fur boots that offered more warmth but less support than leather ones. They could not light the stoves to heat food or melt snow for water. Leaving everything behind and unleashing the dogs, Barne directed the group toward *Discovery,* which he thought lay at anchor only a mile away. "The madness of the act," Scott commented in his diary, "is the more glaring in that he had no knowledge of the country over which he was travelling."[33]

"I impressed on the men, as strongly as I could, the importance of keeping together," Barne later reported. After struggling on the ice, however, one of the men in fur boots, Clarence Hare, abruptly returned to camp for leather ones. When the group fanned out to find him, Edgar Evans slipped off the ridge. Ignoring his own orders to stay together, Barne slid down after Evans, with Arthur Quartly soon to follow. The three landed on a slender cliff edge out of sight and sound of the others. "A yelping dog flew past them, clawing madly at the icy slope, and disappeared forever into the gloom beyond," Scott wrote, based on Barne's report. Hearing nothing from the three men below and fearing the worst, naval seaman Frank Wild led the main group on and soon also reached the precipice. He jumped back from the edge but the other man in fur boots, George Vince, could not stop. "What followed," Scott reported, "was over in an instant. Before his horror-stricken companions had time to think, poor Vince . . . flashed past the leader, and disappeared." Six hours after leaving camp, the main group reached the ship. In three more hours, rescuers found Barne, Evans, and Quartly—all with severe frostbite. Hare stumbled into base two days later; he had become disoriented and slept out the storm under a snowdrift in his gabardine outfit. Scott called it "almost miraculous" that only one man died in the fiasco.[34]

Meanwhile, Royds's advance party encountered problems of their own. Reaching the bare mountainside that blocked their approach to Cape Crozier, they could neither sledge nor ski over the jagged rocks. They could not even climb over them in their leather boots, Royds reported, "as the bruised, sore feet of the party witnessed." After two blinding blizzards, they turned back without posting the ship's location. When the temperature plunged to −42°F one night during their return trip, the three men found themselves unprepared for the cold in their fur night suits. "Sleep proved impossible, and for the first time they found themselves subjected to uncontrollable paroxysms of shivering," Scott wrote in his report. "As the long sleepless hours crept by they had ample opportunities of learning the value of adequate clothing, and the wisdom of being prepared for the unexpected rigours of a fickle climate."[35] One participant, Navy Lieutenant Reginald Skelton, blamed the "girlishness" of the party's leader. He reported telling Scott, "I didn't consider Royds was cut out for sledging; that he is the wrong man for it." Again calling Royds "too girlish" for the work, Skelton privately added that "he certainly has very little brain, and no observing powers at all for scientific or interesting matters."[36] Such gendered comments reflecting masculine views of science and physical stamina, which would sound increasingly shallow to later generations of Britons, were common in Edwardian England. Neither Scott nor Shackleton considered taking women on their expeditions even though some had applied for positions.

To cache supplies for the next year and try his own hand at sledging, Scott dispatched one more overnight trip that first season. Two teams, each with six persons, nine dogs, and two sledges, left the ship on March 31 heading south over the ice. The men and dogs were harnessed together to pull the sledges. They did not get far. "Our loads were arranged theoretically, 200 lbs. to each man and 100 lbs. to each dog, and the first discovery we made was that the dogs entirely refused to work on our theory," Scott wrote. They hung their heads, would not pull steadily with men, liked light loads, and whined piteously. "The fact that the dogs refused to do their share of the work on this trip meant of course that we had to do a good deal more than ours," Scott added. Soon his men were exhausted. The temperature never exceeded −30°F and dropped near −50°F at night. The party struggled with the equipment, especially the fur night suits, which Scott would reconfigure into proper sleeping bags over the winter. "All Fools Day and no doubt we looked it," Seaman James Duncan jotted in his diary on April 1. Scott gave up two days later, deposited the supplies, and returned to the ship after covering only nine miles. It was a sobering experience. Wilson later confided to his diary, "I have *never* realized to such an extent the truth that 'familiarity breeds contempt' as this last year during which I have seen a little of the inside of the 'Royal Navy.' God help it."[37]

Huddled with his officers that winter in the tobacco-smoke fog of the icebound *Discovery*'s wardroom, Scott became his own severest critic. The air could become oppressive. "Although no one smokes out of doors, many smoke within, and a few, amongst whom I must number myself, are inveterate victims of the habit," Scott confessed to his diary. "One turns in every night in a filthy smoking room atmosphere," Wilson, a nonsmoker, complained. "I cannot bless the beastly ass that gave the officers' mess enough cigars to last the winter." Scott brooded on their mistakes. "Our autumn sledging was at an end, and . . . in one way or another each journey had been a failure," he commented. "The errors were patent; food, clothing, everything was wrong, the whole system was bad."[38]

Nevertheless, a certain smugness survived among the officers and crew during that first long winter. After all, they were British, and the British "race," as they called it, still led the way in global exploration and ruled a quarter of the earth's land. Their informal after-dinner discussions and scheduled Tuesday night debates often ranged beyond Antarctic topics to matters of empire and the sacrifices needed to sustain it. For entertainment, Royds organized volunteers into the "Dishcover Minstrel Troupe," complete with blackface, ragged clothes, and heavy dialect. During performances, all joined in the riddling of "Massa Bones" and the singing of African-American songs "with such a will, that the rafters shook," Scott noted. "I don't know why a joke should sound better in nigger language, but I rather think the class of joke made on these occasions does so." Standard fare on the Vaudeville circuit but always repugnant to some, minstrel shows betrayed the cultural views that the *Discovery* explorers carried with them to a white continent. They named their meanest sled dog Nigger, which Scott later complained "wholly failed to convey the grandeur of his nature."[39] The explorers' failures never led them to question their racial superiority.

The lessons of the first fall trips led to better practices during the next summer season, but sledgers still took awful risks, and results fell far below expectations. At the expedition's outset, although it was not stated as an official goal, everyone hoped for the South Pole. Scott did not come close. Scurvy, once again thought eradicated from the Royal Navy after Nares's Arctic expedition, dogged Armitage's Western Sledge Journey and cut short Scott's drive toward the pole at little more than latitude 82° south. To reach even this mark, Scott pushed forward for days on reduced rations against Wilson's medical advice. Shackleton suffered most on this Southern Journey, which by some accounts led to sharp words at the farthest camp and ended with Scott invaliding him home on the first relief ship.

Not much changed during the final summer. Scott's Western Party barely survived a journey over the trackless Polar Plateau without a manual for navigating by the sun. Again, the trek back became a race against death on short rations. That the expedition even had a third season of sledging was due largely to poor planning that kept their ship icebound through the second summer.

Returning to England aboard *Discovery* in September 1904, and prone to self-doubts, Scott feared that he faced the same cold reception that greeted Nares after the failed Arctic campaign of 1875–76. Both expeditions had sent parties closer to a pole than any prior effort but had been stopped far short of that storied destination by inexperience, scurvy, and persistent problems man-hauling heavy sledges—and both had lost at least one man through negligence.

Scott had added grounds for concern. After the RGS's first relief ship, *Morning,* returned in 1903 with news that *Discovery* was icebound and scurvy had reappeared, Markham feared a repeat of the Franklin disaster if the expedition did not get out in 1904. Knowing the RGS could not afford a second relief effort and foreseeing a need for two ships to evacuate Scott's party, Markham asked the government for aid. When it balked, Markham openly accused Prime Minister Arthur Balfour of letting a naval expedition perish in the field to save a few thousand pounds. Balfour responded by telling the House of Commons that he had lost confidence in the RGS and would support the relief effort provided the Admiralty handled the entire operation, took control of *Morning,* and brought everyone back that year with or without Scott's ship. Markham's overreaction had transformed a standard relief operation into an emergency rescue mission.

Resigned to a third winter in the ice and feeling that the expedition was in no danger, Scott despaired when two ships, *Morning* and *Terra Nova,* arrived with orders for him to abandon *Discovery* if it could not sail north that year. He later wrote that "this mandate placed me and my companions unavoidably in a very cruel position."[40] Accepting rescue meant admitting failure. Even after *Discovery* broke free of the ice at the last minute, Scott feared he might face criticism for the costly relief effort. Upon reaching New Zealand, he opened himself to further censure by telling the press that the Admiralty had wasted money by sending two vessels to save him. A generation earlier, Markham had tried to make the most of his cousin's Farthest

North but failed to persuade the public that Nares's expedition was anything but a flop. Now, however, the British people wanted heroes and saw one in Scott. At least until World War I broke the spell, Markham's Arthurian mindset had come back into vogue with an Antarctic Camelot and the South Pole as its Holy Grail.

The empire warmly welcomed the explorers. Celebrations begin in April 1904, when the returning ships reached their first port, Lyttelton, New Zealand, to find townspeople lining the harbor to greet them. Banquets, balls, and receptions followed. While *Discovery* was made ready for the voyage home, Scott polished his speaking skills with public lectures in sold-out halls throughout the colony. "New Zealand," he wrote, "welcomed us as its own."[41]

The process repeated itself on a grand scale when *Discovery* arrived in England five months later. Members of the expedition received an exuberant homecoming when they docked in Portsmouth, where the mayor hailed their "British grit," and three days later in London, where they were met by officials of the City, Admiralty, Royal Society, and RGS. "That their endeavors were recognized by the nation was strongly exemplified by the actions of the King," a Royal Society officer stated on this occasion. "His Majesty had felt the nation's pulse, and had prescribed an Antarctic medal" for each of them. This medal, a top admiral added, "was not inferior in its significance to any war medal."[42] Indeed, for many, Scott and his men became the national heroes that the Boer War had failed to supply. Markham, speaking for the RGS, praised the expedition as "an achievement unequalled, certainly unsurpassed, in the annals of polar exploration—Arctic or Antarctic." In an earlier speech at the RGS, he had already asked, "Is there any tale of derring-doe surpassing the story of those who have planted the cross of St. George in 82° 17′ S.?" The *Times* decreed the entire venture "an almost unqualified success."[43]

The explorers became celebrities. Crowds packed their public appearances. In November, an exhibit of Wilson's Antarctic art attracted over ten thousand viewers to London's prestigious Bruton Gallery. Advertisements in the gallery catalogue touted paints and binoculars used by the artist in the Antarctic. Sensing the expedition's mass appeal, other suppliers followed suit with advertisements of their own noting the explorers' use of Cadbury chocolate, Colman's mustard, Burberry gabardine clothing, Ideal typewriters, Primus paraffin stoves, and Burroughs Wellcome medical equipment. "I have much pleasure in recommending Crease's White Petroleum Jelly," Scott's partner on the second Western Sledge Journey, Edgar Evans, stated in an ad. "In cold climates it is invaluable for use after shaving." Another ad quoted Scott hailing Bovril beef extract as excellent "on sledge journeys" and "especially useful with the cooking of seal meat."[44]

Within this new constellation of polar stars, Scott's light blazed brightest. King Edward VII made him a commander in the Royal Victorian Order, one step below knighthood, and invited him to the royal residence at Balmoral Castle for four days of dining, dancing, and hunting. Other invitations poured in as Scott became the Lion of the Season in London society. Despite his lacking the requisite time at sea, the Admiralty promoted him to captain and released him to write his account of the expedition. Publishers vied to bring out the two-volume narrative, which netted Scott more than four years of navy pay. Seven thousand people attended his address to the RGS in the Royal Albert Hall, where he declared himself "overcome by the rain of medals." He received two that evening, with five more to come from European and American geographical societies. Months of speeches ensued as Scott traveled the lecture circuit, giving the same talk in dozens of British towns and averaging more income from each than he earned for a month of navy service. Everywhere, people wanted to hear about his "Farthest South," as he titled the lecture, with the

American ambassador proclaiming at the RGS address that if Britain let Scott return south to "complete" his work and America granted a similar opportunity to Robert Peary in the North, then both would reach a pole and "leave the globe we inhabit, as it should be, in the warm and fraternal embrace of the Anglo-Saxon race."[45]

For many, Scott came to represent ideals of British manhood in the Edwardian age. "Neither tall nor short," one journalist noted, Scott "conveyed the impression of strength and vigor, . . . with something about it which told you that the man was used to speaking the crisp language of command." Markham hailed him as "alike the skillful and bold navigator, the ideal director of a scientific staff, the organizer of measures securing the health and good spirits of his people, and the beloved commander of [a] chosen band." The *Times* wrote, "The example of 'grit' and self-sacrifice in the service of science shown by Captain Scott and his companions will, no doubt, serve to keep alive among us that spirit which has done so much for England's greatness in the past." Others spoke of his "efficiency," "skill and courage in the most difficult circumstances," and "patient perseverance." He epitomized every good trait of the national character that pundits saw slipping away in the anxious years following the Boer War.[46]

In praising Scott and his men, British commentators typically focused on their rigorous sledge journeys and contributions to science even though sledging had advanced further in the Arctic and the expedition's scientific findings were preliminary at best. Their comments reflected the values of the Edwardian age: fitness and science mattered. Indeed, for those like Galton, physical and mental fitness had become a product of eugenic science, and doing good science a sign of moral and national fitness. In some respects, the expedition itself represented a scientific experiment in building better Britons through selecting the best stock and subjecting it to a struggle for existence in the world's harshest environment. It seemed only right that the explorers read Darwin on the sledging trips and that Scott titled his book *Voyage of the "Discovery"* after Darwin's *Voyage of the Beagle*. For decades, Markham had described geographical exploration as a way of measuring the world, the empire, and one's own manliness. Scott and his men seemingly accomplished all three. "They," Markham said on their return, "would always look back to the recent period of their lives, and feel that they had had opportunities to make them better and finer men."[47] By his standards, they had proved themselves fit.

Shackleton was the lone exception. Although cordially received on his early return, he felt that Scott had disgraced him first by declaring him unfit to remain a second year and then by depicting him in *Voyage of the "Discovery"* as borne back from their Farthest South on a sledge—a point Shackleton vigorously denied. He sought to gain redemption by mounting his own Antarctic expedition. "Men go out into the void spaces of the world for various reasons," he wrote to open his account of that venture. "I had been invalided home before the conclusion of the *Discovery* expedition, and I had a very keen desire to see more of the vast continent that lies amid the Antarctic snows and glaciers." And, Shackleton could have added, to prove himself. *Discovery*'s doctor had diagnosed him as asthmatic, a potentially incapacitating condition. "The wheezing in his lungs, a tightness in his chest, were not just an affront to his manhood," biographer Roland Huntford wrote of Shackleton's humiliation on leaving *Discovery*, "they were flaws threatening fulfillment." He would fight back. Reaching the South Pole, which Scott had fallen short of doing, became Shackleton's stated goal. If competition marked the Edwardian age, then he personified it. "He told me he had been planning something of the kind ever since he came back," RGS Secretary John Scott Keltie wrote to Scott about Shackleton early in 1907, "probably to prove that though he had been sent home, he is quite as good as those who remained."[48]

Compared with the *Discovery* mission and Scott's as yet unannounced second effort, the new expedition came together quickly. Following the heroes' welcome given to *Discovery*'s returning explorers, volunteers flocked to join Shackleton: more than four hundred applicants for fifteen slots on the shore party. Despite his lack of interest in science, Shackleton assembled a superior scientific staff that could better Scott's showing. With his infectious enthusiasm winning financial backers and publishing contracts, he secured sufficient support for a barebones expedition in a matter of months and set off on the converted sealer *Nimrod* in August 1907. Eschewing the naval discipline that marked the *Discovery* expedition, Shackleton's charismatic personality provided all the authority needed to maintain order and effective command. Those who knew him worried about his fitness. "He looks strong enough," Keltie wrote to Scott, but "he is not absolutely sound, and Heaven knows what may happen if he starts on his journey Pole-ward." Shackleton's wife insisted that he see a heart specialist before departing, but according to a doctor familiar with the case, Shackleton never let the specialist listen to his heart for fear "that he wouldn't be fit to go to the Ross Sea."[49]

Shackleton outlined his strategy for reaching the South Pole in his initial announcement to the press. "It is held that the southern sledge party of the *Discovery* would have reached a much higher latitude if they had been more adequately equipped for sledge work," the *Times* reported, "and in the new expedition, in addition to dogs, Siberian ponies will be taken as the surface of the land or ice over which they will have to travel will be eminently suited for this mode of sledge travelling." To please the expedition's principal backer, who owned a motor works, Shackleton also took along an automobile fitted with sleigh runners in front and boasted that it could cover 150 miles per day over the Ice Barrier. "He thinks there would be a fair chance of sprinting to the Pole, though we fancy his estimated daily average is far too great," England's leading car magazine commented.[50] Clearly Shackleton anticipated man-hauling the sledges in the end, especially if, as he suspected, the pole lay on the high plateau beyond the mountains of South Victoria Land. He never attempted to train the dogs suitably for long-haul sledging and could not count on either ponies or a motor car to scale the mountains and travel at altitude. Realistically, they would mainly help transport the polar party across the Ice Barrier, with the ponies also serving as fresh meat for the scurvy-wary explorers.

Antarctic conditions derailed Shackleton's innovative plans for polar transport. Unable to operate beyond the smooth sea ice near the expedition's Cape Royds base, the car contributed little to the drive across the Ice Barrier. The ponies, though they pulled better than *Discovery*'s dogs, proved ill-suited for Antarctica's deep snow and frigid winds. They required excessive amounts of fodder, which greatly added to the load. Sweating through their hides, they froze easily on the march and required special care when stopped. Worst of all, with their great bulk and small hoofs they frequently broke through the crust on the Ice Barrier, which made for slow going in snow up to their haunches, and occasionally fell through ice bridges over deep crevasses, which finally claimed the last of them on the long haul up Beardmore Glacier to the Polar Plateau. After that it was back to man-hauling, which was perhaps what Shackleton wanted all along because only that could demonstrate the fitness of men to British standards and result in a story with more daring and drama than Scott's Farthest South. Fighting bitterly cold winds at high altitude on scant rations, the party struggled to within 112 miles of the pole before turning back for another horrific race-against-death back to base. "Our geological work and meteorology will be of the greatest use to science," Shackleton wrote in his diary near the end of his southbound march, "but all this is not the Pole. Man can only do his best, and we have arrayed against us the strongest forces of nature."[51]

His best was good enough for Britons. He had told his wife before he left, "I am representing 400 million British subjects," and upon his return, they adopted him. For a season his star shone brighter than Scott's. Shackleton had sold first rights to news of his return to London's *Daily Mail,* which he honored by secretly wiring the story from a small island off the New Zealand coast before sailing into Lyttelton Harbor to a tumultuous reception on March 25, 1909, less than two weeks before Peary laid claim to reaching the North Pole. *Pearson's Magazine* had rights to a serialized version of the explorer's narrative, London publisher William Heinemann held the book rights, and Christy's Lecture Agency would handle the speaking tour. All had reason to promote the story, which quickly became one of imperial triumph. A Sydney journal, noting the role of Edgeworth David, Douglas Mawson, and other Anglo-Australians in the story, called it "a tale of the race." Embittered that Shackleton had displaced Scott in the pantheon of polar explorers, Markham complained, "The exaggerated praise of cleverly engineered newspaper articles raised him to the crest of a wave . . . above his chief," but the adulation represented something deeper than fabricated fancy.[52]

The response was greatest in Britain, where it reached even into Markham's RGS. "I don't remember anything like it since the return of Nansen," Keltie wrote to Shackleton in April. "An amount of pluck and determination has been displayed by Lieut. Shackleton and his companions which has never been surpassed in the history of Polar enterprise," the RGS's *Journal* declared in announcing the achievement. "When we are all feeling a little downhearted at seeing our supremacy in sport and in more serious matters slipping away from us," one London newspaper commented, "it is a moral tonic to find that in exploration we are still the kings of the world." Another added,

"So long as Englishmen are prepared to do this kind of thing, I do not think we need lie awake at night dreading the hostile advance of 'the boys of the dachshund breed.'" Based on such accounts, historian Beau Riffenburgh described Shackleton as "the elixir that the Empire needed for the self-doubts brought about by the debacle of the South African War, for concerns about the thunderclouds slowly forming over Europe, and for the loss of economic and physical supremacy."[53]

After touring New Zealand and Australia, Shackleton sped to London by steamship and railway to claim his laurels while *Nimrod* took a slow route home exploring the Southern Ocean. New Zealand journalist Edward Saunders accompanied him to ghostwrite the expedition narrative, *The Heart of the Antarctic,* from the explorer's dictated bursts so that it could appear before public interest flagged. A cheering crowd of over five hundred greeted Shackleton at London's Charing Cross Station on June 15, and two weeks later he addressed a packed house at the Royal Albert Hall, where he received the RGS's gold medal from the Prince of Wales and praise from the society's new president, Charles Darwin's son Leonard, a former explorer and prominent eugenicist. Among the many honors of empire that followed, the king raised Shackleton initially to Scott's ceremonial rank of commander in the Royal Victorian Order and then, as the public response kept building, above Scott with a knighthood. Parliament voted £20,000 to pay off expedition debts. "We seem to be living in times when men have reverted to the age of the elemental heroes," one magazine observed in late summer. An English teachers' journal concluded that Shackleton's "achievement is one that appeals to all sorts and conditions of people, from the small school-girl, who is always looking out for a hero to worship, to the staid geographer who objects to blanks on maps."[54]

* * *

Now it was Scott's turn to feel excluded. Like Markham, he resented Shackleton's success and the acclaim it engendered. "Shackleton owes everything to me as you know," he had written to Keltie on first learning of his rival's plans, and he felt betrayed by a former subordinate attempting to beat him to the pole using the route that he had charted. After Shackleton agreed under pressure not to use Scott's route but then did so when others appeared infeasible, Scott called him "a professed liar" and "plausible rogue." Prodded by RGS leaders, he joined the official party welcoming Shackleton back to London and followed the party line by hailing the new Farthest South as "a glorious example of British pluck and endurance." Yet Scott did not believe that Shackleton had gotten as close to the South Pole as claimed, and he used the homecoming banquet to announce his own willingness to try again for the pole. He would do so, he declared, "for the honor of the country." In fact, he had been planning such an effort since before *Nimrod* sailed, and on hearing that Shackleton had fallen short, he said to a *Discovery* colleague, "I think we had better have a shot next."[55]

Scott announced his new expedition in September 1909 and set his departure for the following summer. While promising a spirited scientific program, he bluntly declared planting the Union Jack at the South Pole as his main goal, asserting that it would show the nation's greatness. The *Times* story on Scott's announcement spoke of British blood "being stirred by patriotic emulation in a noble field." Sailing without RGS funding, Scott appealed to the public for money. The British had showed the way to the pole, he implored, but foreign teams would get there first unless the British quickly returned to the field. "There is one Pole left, and that should be our Pole," the writer Arthur Conan Doyle declared in support of Scott's plea.[56] In the euphoria surrounding the return of Peary and Shackleton, over eight thousand applications poured in for thirty-two places in the shore party. Money came too, from £20,000 voted by Parliament to small sums sent by school groups to "sponsor" dogs, ponies, and even Scott's sleeping bag. "Captain Scott is going to prove once again that the manhood of the nation is not dead," RGS President Leonard Darwin stated at the departure luncheon. "To my mind, this is the most solid asset to be set down to the credit of this expedition." In a fitting reply, Scott said, "I feel that we have gathered together a set of men who will represent the hardihood and the energy of our race."[57] They sailed on *Discovery*'s old rescue ship, *Terra Nova*.

Despite a complex initial plan to use tractors, ponies, and dogs to place a party at the pole, in the end Scott relied on manpower. The three motorized tractors, though tested in Norway, proved little help beyond the sea ice and gave trouble even there. One broke through the ice when being offloaded from the ship, and the other two gave out during the first week of the polar trek. "The motors are not powerful enough to pull the loads over heavy surfaces as they are constantly overheating," driver William Lashly complained three days into the journey. "The distance in each run is about a thousand yards to a mile. Then it is necessary to stop."[58] The ponies outperformed the tractors but created almost as much work as they saved. Chosen in Siberia less for their overall condition than for their white color, which Scott saw as a sign of fitness for Antarctica, they struggled in laying advance depots on the Ice Barrier during the first season and ultimately placed the farthest one thirty-six miles short of the original goal, handicapping the next year's polar effort. Even so, six of the eight ponies on the late autumn depot journey died, leaving only ten for the polar trek that began in November. Caked with ice in blizzards, out of food, sunk to their bellies in soft snow, and slowing the entire effort by their pace, the last surviving ponies were put down just short of Beardmore Glacier.

The dogs performed better than Scott expected, but he would not rely on them for ascending Beardmore Glacier or crossing the Polar Plateau. According to plan, after reaching the glacier he sent them back with their sledges and handlers to relay supplies for the return journey. His party would man-haul the remaining 360 miles up the glacier and across the plateau to the pole, while Roald Amundsen's team sped ahead with dogs on its converging path to glory.

For a century, polar experts and Antarctic historians have questioned Scott's judgment in choosing man-hauling over dog-sledging in the face of compelling contrary testimony from the work and words of Nansen, Peary, and a generation of Arctic explorers. Scott's bad experiences with ill-trained dogs on the *Discovery* expedition and his sensitivity to canine suffering undoubtedly influenced his decision, but to a degree, virtually every answer to the question of man-hauling comes back to Edwardian measures of manliness and competitive demonstrations of fitness. "In my mind," Scott famously wrote in *Voyage of the "Discovery,"* "no journey ever made with dogs can approach the height of that fine conception which is realized when a party of men go forth to face hardship, dangers, and difficulties with their own unaided efforts, and by days and weeks of hard physical labour succeed in solving some problem of the great unknown. Surely in this case the conquest is more nobly and splendidly won."[59] To prove "the manhood of the nation," as Leonard Darwin had termed it, Scott and his men had to pull their sledges to the pole and back or die trying.

By the time of his second expedition, man-hauling was an established feature of British polar exploration. It expanded with the Franklin searches and was romanticized by some veterans of that effort, including Clements Markham. He included in the *Discovery* expedition's manual an essay on sledging by Leopold McClintock, who wrote of the searches, "The endurance of the hardiest was called forth, and the talent of invention evoked and stimulated, until at length a system of sledging was elaborated." This system relied on man-hauling, and the talent it evoked became part of Victorian lore that Markham, Scott, and Shackleton carried into a new century. In planning the *Discovery* expedition, for example, after unfavorably contrasting the efforts of Nansen and Peary using skis and dogs "with that work done by British seamen" man-hauling sledges, Markham declared: "That is the way to travel in polar regions. No *ski,* no dogs." As for sledging with dogs in the Arctic, he wrote in 1899, "Nothing has been done with them to be compared with what men have achieved without dogs." And to honor his role in engineering that expedition, when its members returned they gave Markham a statuette of a man pulling a sledge on foot. Pelham Aldrich, who commanded the grueling Western Sledge Journey during Nares's Arctic expedition, wrote warmly to Scott in 1903, "Your sledging will be handed down to posterity as being fully worthy of a foremost place in Arctic or Antarctic work generally."[60] After Shackleton carried the tradition almost to the South Pole and back to great national acclaim in 1909, Scott continued it on his return to the Antarctic.

Members of the *Terra Nova* expedition by and large embraced the ethos of man-hauling. During a midwinter lecture in the Antarctic, Scott explained his preference for relying on ponies and men to reach the pole, with ponies used for the Ice Barrier and men beyond. "With this sentiment the whole company appeared to be in sympathy," he reported. "Everyone seems to distrust the dogs when it comes to glacier and summit." When the tractors broke down early on the polar journey, Lashly seemed almost to welcome it, writing in his diary, "Now comes the man-hauling part of the show." On seeing the broken tractors, Scott exclaimed, "So the dream of great help from the machines is at an end!" When the ponies gave way next, Wilson proclaimed, "Thank God the horses are now all done with and we begin the heavier work ourselves." Scott marked the departure of the dogs by noting that soft snow on Beardmore Glacier made hauling sledges "very difficult for dogs," but with the load transferred to men, "I found we could make fairly good headway." The final support party turned back after reaching the Polar Plateau, leaving Scott and four others to cover 150 miles to the pole hauling one sledge. In a letter sent back to Kathleen Scott with the support party, H. R. "Birdie" Bowers—who faced the journey on foot while the others had skis—wrote, "After all it will be a fine thing to do that plateau with manhaulage in these days of the supposed decadence of the British Race."[61]

Discovery, Nimrod, and *Terra Nova* sailed in the fleeting period following the fin de siècle when many Europeans saw meaning in heroic struggle against fate and seemingly futile displays of human strength, before the Great War made such efforts appear hollow. Having attained individual recognition through their actions on the ice, many veterans of the Antarctic expeditions served in a war in which more than a million members of the British Imperial Forces died, most of them among the undifferentiated masses mowed down in its trenches and battlefields. A few of the explorers themselves died, mostly in action at sea. For supporters of polar exploration like Markham and Leonard Darwin, who succeeded Francis Galton to chair Britain's Eugenics Education Society in 1911, the expeditions served as an experiment in measuring racial fitness and character—notions that were sorely tested in the Great War but hung on at least until a second world war.

The participants shared at least some of these views. They were part of the culture. "I do not think there can be any life quite so demonstrative of character as that which we had on these expeditions," Scott wrote from the Antarctic in 1911. "Here the outward show is nothing, it is the inward purpose that counts. So the 'gods' dwindle and the humble supplant them. Pretense is useless." Speaking of the explorers' scientific fieldwork, he later added, "The achievement of a great result by patient work is the best possible object-lesson for struggling humanity."[62] Even as the explorers tested themselves by reaching for the pole, they attempted to give added meaning to the expeditions through scientific programs that, as Antarctica became better known, gradually shifted focus from terrestrial magnetism, geographical discovery, oceanography, and meteorology to biology, geology, and glaciology.

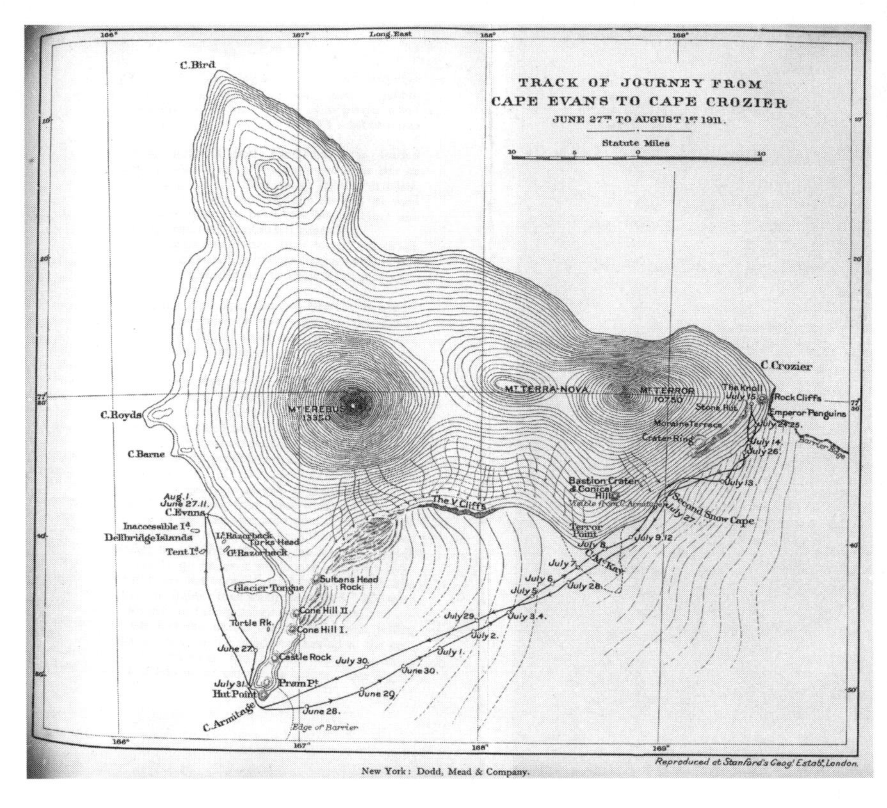

Route of the winter journey of Edward Wilson, Birdie Bowers, and Apsley Cherry-Garrard around Ross Island to Cape Crozier (1911), from *Scott's Last Expedition* (New York, 1913).

CHAPTER 6

March to the Penguins

DISCOVERY SAILED FOR SCIENCE. ALL INVOLVED said as much. One of the expedition's two sponsors was the Royal Society, the British Empire's leading voice for basic research in science. In discussing government support for the venture, First Lord of the Treasury Arthur Balfour commented, "It would not be creditable to an age which flatters itself, above all other ages, to be a scientific age, if without reluctance we acquiesced in the total ignorance which now envelopes us of so enormous a portion of the southern hemisphere." King Edward VII, in his sendoff speech from *Discovery*'s deck, told the explorers, "I have often visited ships in order to say farewell when departing on warlike service; but you are starting on a mission of peace, and for the advance of knowledge."[1]

Some expedition proponents and participants clearly had motives that had little to do with science. Scott hoped for promotion and perhaps the pole; Markham dreamed about deeds of "derring doe"; the British public hungered for heroes; the RGS wanted to fill blank spaces on the world's maps. None of these goals precluded science. In Edwardian England, doing good research in extreme conditions was a reason for promotion, a recognized species of adventure, and a significant part of what made the effort heroic. Without gathering facts and collecting specimens from nature along the way, the explorers simply could not fill the globe's largest blank space to the RGS's satisfaction. Recounting the results of the expedition's first year while *Discovery* remained locked in the Antarctic ice, Markham tellingly exclaimed to RGS members, "Think of them now, entering cheerfully upon a second winter! Think of the terrible hardships and suffering they have gone through for science!"[2]

To gain the government, institutional, and public support needed to fund the expedition on the scale they wanted, Markham and other RGS leaders added various scientific justifications to what they originally envisioned as mainly a daring journey of geographical discovery. Promises of research in terrestrial magnetism, meteorology, oceanography (including marine biology), and later in geology and glaciology all found their way into the expedition's official instructions. When it sailed, *Discovery* carried all manner of magnetic instruments, weather gauges, drift nets, and dredging equipment. Given the intense subsequent interest in Antarctic biology and the raging early-twentieth-century debates over evolutionary mechanisms, however, the initial discussion and directions for the expedition made surprisingly few references to the study of animals and plants on land. A prospectus of *Discovery*'s scientific program published in the science journal *Nature* explained, "The biological work of the expedition will be mainly at sea; for the ancient maxim that 'Nature loves life' does not appear to apply to the Antarctic lands. The main biological duty of the expedition is to make as extensive a collection of the fauna and flora of the Antarctic Ocean as the ship's storage will admit."[3]

The omission of shore-based biology from *Discovery*'s agenda was all the more remarkable because naturalists and officers on British naval expeditions from the days of James Cook onward had proved adept at collecting animals and plant specimens on land. It neatly fit the natural-history emphasis of nineteenth-century British science. The RGS's *Hints to Travellers,* which Victorian era British explorers and intrepid tourists carried to the far corners of the earth, included

HMS *Erebus* and *Terror,* the first ships to sail through the Antarctic ice pack and enter the Ross Sea, with Mount Erebus erupting in the background, January 1841, from James Ross's *Voyage of Discovery* (London, 1847).

Clements Markham, 1905, with the statuette of an Antarctic explorer hauling a loaded sledge given to him by the members of the *Discovery* expedition, from Robert Scott's *Voyage of the "Discovery"* (New York, 1905).

Robert Falcon Scott in his Royal Navy uniform, 1905, as pictured in the frontispiece of Scott's *Voyage of the "Discovery"* (New York, 1905).

Ernest Shackleton, 1909, as pictured in the frontispiece of Shackleton's *Heart of the Antarctic* (Philadelphia, 1909).

Ernest Shackleton, Robert Scott, and Edward Wilson (left to right) preparing to leave for the Southern Sledge Journey across the Great Ice Barrier in November 1902, from Robert Scott's *Voyage of the "Discovery"* (New York, 1905).

Winter quarters for the *Discovery* expedition, 1902–4, showing the ship in the background and the snow-covered instrument huts for scientific research in the foreground, from Robert Scott's *Voyage of the "Discovery"* (New York, 1905).

Charles Royds, who collected weather data for the *Discovery* expedition, shown with the expedition's land-based meteorological instruments at Hut Point with the ship *Discovery* in the background, from Robert Scott's *Voyage of the "Discovery"* (New York, 1905).

Sledging with men and dogs in heavy snow early in the *Discovery* expedition before Scott learned that men and dogs could not effectively pull together, from Robert Scott's *Voyage of the "Discovery"* (New York, 1905).

Louis Bernacchi, who conducted magnetic research on both the *Southern Cross* and *Discovery* expeditions, exiting the hut containing instruments to measure terrestrial magnetism at the *Discovery* expedition's winter quarters, from Robert Scott's *Voyage of the "Discovery"* (New York, 1905).

Entertainment at the *Discovery* expedition's winter quarters showing members of the expedition dressed in blackface for a comedic minstrel show, which reflected the eugenic racial views that helped to spur British overseas exploration and colonization during the late nineteenth and early twentieth centuries, from Robert Scott's *Voyage of the "Discovery"* (New York, 1905).

Nimrod held up by pack ice (1908), from Ernest Shackleton's *Heart of the Antarctic* (Philadelphia, 1909).

James Murray and Raymond Priestley descending the shaft cut into a frozen lake near the *Nimrod* expedition's winter quarters to study the composition of the ice and contents of the lake, from Ernest Shackleton's *Heart of the Antarctic* (Philadelphia, 1909).

Members of the *Nimrod* expedition in February 1909 excavating a six-year-old depot from the *Discovery* expedition to measure glacial movement and snowfall accumulation on the Great Ice Barrier, from Ernest Shackleton's *Heart of the Antarctic* (Philadelphia, 1909).

Blocks of lake ice hanging from the *Nimrod* expedition's hut at Cape Royds to research the ablation of ice due to the evaporation of ice vapor during the harsh Antarctic winter, from Ernest Shackleton's *Heart of the Antarctic* (Philadelphia, 1909).

Ernest Shackleton, Frank Wild, and Jameson Adams (left to right), in a photograph taken by Eric Marshall, at the flag marking their Farthest South on the Polar Plateau, January 9, 1909, from Shackleton's *Heart of the Antarctic* (Philadelphia, 1909).

Terra Nova in the pack ice during the second expedition led by Robert Scott to the Antarctic (1910), from Apsley Cherry-Garrard's *Worst Journey in the World* (New York, 1922).

Raymond Priestley with an erratic granite boulder at Cape Royds, which he saw as evidence of former, more extensive glaciation in Antarctica, from Ernest Shackleton's *Heart of the Antarctic* (Philadelphia, 1909).

Forbes Mackay, Edgeworth David, and Douglas Mawson (left to right) at the flag marking the South Magnetic Pole, January 16, 1909, from Ernest Shackleton's *Heart of the Antarctic* (Philadelphia, 1909).

Kathleen Scott, Robert Scott, and expedition business manager G. F. Wyatt (left to right) on *Terra Nova*'s bridge before the ship's departure from Port Chalmers, New Zealand, for Antarctica, on November 29, 1910. © Scott Polar Research Institute Archives.

Robert Forde, Frank Debenham, Tryggve Gran, and Griffith Taylor (left to right) on the second Western Geological Journey of the *Terra Nova* expedition, December 25, 1911, from *Scott's Last Expedition* (New York, 1913).

Edward Wilson and Birdie Bowers (left to right) recording the temperature at Cape Evans in midwinter darkness, 1911, during the *Terra Nova* expedition, from *Scott's Last Expedition* (New York, 1913).

Frank Debenham, Apsley Cherry-Gerrard, Birdie Bowers, Teddy Evans, and Griffith Taylor (left to right) working at the *Terra Nova* expedition's winter quarters in 1911, from *Scott's Last Expedition* (New York, 1913).

Birdie Bowers, Edward Wilson, and Apsley Cherry-Garrard (left to right) preparing to leave for their winter journey to study emperor penguins at Cape Crozier in June 1911, from *Scott's Last Expedition* (New York, 1913).

The Northern Party—George Abbott, Victor Campbell, Harry Dickason, Raymond Priestley, Murray Levick, and Frank Browning (left to right)—in September 1912 after winter in a snow cave, during the *Terra Nova* expedition. © Scott Polar Research Institute Archives.

Teddy Evans surveying with a theodolite during the *Terra Nova* expedition, from *Scott's Last Expedition* (New York, 1913).

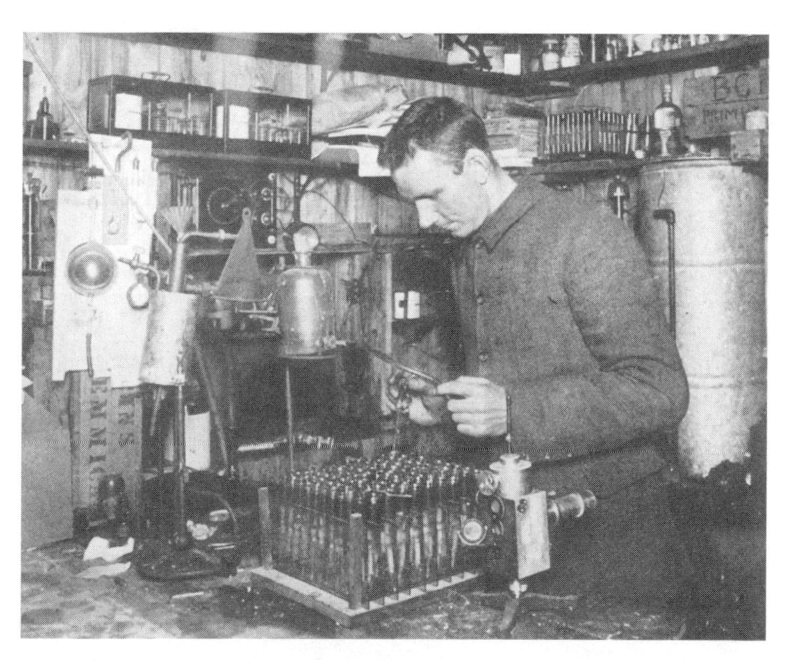

Meteorologist George Simpson conducting laboratory research at winter quarters during the *Terra Nova* expedition, from *Scott's Last Expedition* (New York, 1913).

Physicist Charles Wright taking observations with the transit in midwinter during the *Terra Nova* expedition, from *Scott's Last Expedition* (New York, 1913).

Roald Amundsen in Inuit Arctic clothing, his "polar suit," as pictured in the frontispiece of Amundsen's *South Pole* (New York, 1913).

Roald Amundsen's ship *Fram* at the Great Ice Barrier in January 1912, from Amundsen's *South Pole* (New York, 1913).

Norwegian explorers using a theodolite to determine the location of the South Pole, from Roald Amundsen's *South Pole* (New York, 1913).

Norwegian explorers, with Roald Amundsen at far left, at the tent left to mark the South Pole, from Amundsen's *South Pole* (New York, 1913).

Titus Oates, Birdie Bowers, Robert Scott, Edward Wilson, and Edgar Evans (left to right) at the South Pole, January 18, 1912, from *Scott's Last Expedition* (New York, 1913).

in its practical travel advice detailed directions on skinning animals, stripping skeletons, preserving invertebrates, and pressing plants for museum collections. Perhaps expedition organizers simply assumed that British naturalists and officers on a voyage of exploration would collect land animals and plants without being told; and of course the explorers would carry *Hints to Travellers*. Still, much of the material on zoology in the *Antarctic Manual* prepared by the RGS for the *Discovery* expedition dealt with fish, whales, and deep-sea life, and its brief chapter on botany began by noting "It is very improbable that any flowering plant will be found in the Antarctic regions to be visited by the Expedition," before focusing on how to collect and preserve seaweed, plankton, and other marine plants.[4]

Nevertheless, the organizers knew that Victoria Land and Ross Island provided a rich habitat for seals and seabirds, particularly penguins, petrels, and skuas. At the time, European zoologists knew little about the region's Weddell and Ross seals and nothing about the breeding habits of emperor penguins. And European and American collections at the time typically contained too few specimens of any Antarctic seal or seabird species to study its variation, distribution, or evolutionary development. In an era when biologists hotly debated how evolution operated and many nonscientists questioned whether it happened at all, penguins held particular interest as either a surviving link between ancient flightless forms and modern birds or an evolutionary example of birds that lost the seemingly advantageous ability to fly. Seals had commercial value.

Although the *Antarctic Manual* duly urged the explorers to look for the breeding ground of emperor penguins and to collect seabirds and seals, the initial discussions and instructions for the expedition show a striking lack of attention to such matters. At scientific planning meetings, for example, the Zoological Society's perennial secretary P. L. Sclater dismissed Antarctica as not "particularly favorable for zoological work on the higher land vertebrates," while evolutionary biologists D'Arcy W. Thompson and Ray Lancaster, both later knighted for their work in science, stressed only the mission's value for marine research. Even the revered senior botanist Joseph Hooker, the last surviving member of Ross's Antarctic expedition of 1839–43 and of Charles Darwin's inner circle, spoke solely about oceanic plant life when promoting the expedition to government officials. "We know there is no terrestrial vegetation there," he advised Balfour.[5] Perhaps reflecting the relative priority of these activities, Scott put the expedition's official biologist, Thomas Hodgson, in charge of studying fish and marine invertebrates; asked its chief surgeon, Reginald Koettlitz, to serve as botanist with the duty of reporting on marine microorganisms; and turned over the task of collecting seals and seabirds to the assistant surgeon, Edward Wilson, who also acted as expedition artist.

Wilson applied himself to the project with characteristic zeal. Son of a physician, Wilson studied medicine and science at Cambridge, where he adopted the idiosyncratic, quasi-mystical Christian faith that marked his later life. He attended Protestant services but frequently took communion alone, contemplated the lives of Catholic saints, and embraced asceticism. His faith did not cause him to deny the theory of evolution, but it led him to see God as active in the evolutionary process. While training to become a doctor, Wilson chose to work and live at an evangelical mission in the slums of London, where he contracted a tuberculosis-like ailment that nearly killed him. He developed his skills as a scientific and medical illustrator during his recuperation, attracting the attention of the ornithologist P. L. Sclater, a family friend, who recommended him as a surgeon and artist for the *Discovery* expedition.

At first, Wilson feared that his illness would disqualify him, but the imperious Sclater brushed aside such concerns. "I am on the committee that makes the appointments," he assured Wilson, "and my opinion is that you would be a suitable person for the post." Wilson failed the medical exam for the expedition but was authorized to go at his own risk. Accustomed to viewing hardships as gifts from God for self-improvement, he wanted nothing more than this. "If the climate suits me," the young doctor wrote matter-of-factly, "I shall come back more fit for work than ever, whereas if it doesn't I think there is no fear of me coming back at all. I quite realize that it is kill or cure." The position offered Wilson a seemingly God-given opportunity to combine his passions for medicine, science, art, and asceticism. He married a like-minded Christian, Oriana Souper, three weeks before *Discovery* sailed, and together they practiced three years of marital separation. "Three weeks with my Ory is food enough for three years' hope," Wilson wrote in his diary on New Year's Eve, 1901. "God keep her."[6]

Two brief, unscheduled stopovers on the voyage south demonstrated the eagerness of *Discovery*'s officers and naturalists to do botany and zoology in the manner of Royal Navy expeditions. Both were at uninhabited oceanic islands that the explorers saw as near-virgin territory for collecting. They scoured them like boys on a scavenger hunt.

The first stop came on the way to Cape Town at the tropical South Atlantic island of Trindade, then claimed by both Britain and Brazil. "Few naturalists have landed on it," Scott noted, "and as it lay on our route I thought our time would not be wasted in giving our officers an opportunity of a run ashore." They anchored the ship off the rocky coast and took whaleboats to shore. "All of us scientists embarked with Royds as captain of my boat and the Skipper of the other," Wilson wrote. "Our orders were to collect and bring on board everything of natural history interest that we could lay our hands on. We were strung up like Christmas trees with bags and satchels and boxes and bottles." Six hours later, they returned to the ship with bags of eggs, satchels of birds, boxes of plants, bottles of insects, a fifteen-foot tree fern, and Wilson's live-animal sketches. "Shackle found grasshoppers, cockroaches, ants, also a gannet's nest made of sticks," Wilson reported. "Skelton shot many white breasted petrels, which he and I found. . . . I took one egg off a bare rock—no nest at all. The tern was sitting on it and, poor bird, I left her sitting still on the same spot when I left."[7] They claimed to discover an unknown species of petrel, which they named for Wilson, but ornithologists later demoted it to a subspecies. Scott remembered the excursion as a satisfactory first outing for science.

Ten weeks later, on the way from Cape Town to New Zealand, *Discovery* drew abreast of another rocky outcropping: Macquarie Island, far south of Australia. Scott did not plan to stop, but Wilson bribed the ship's pilot with a bottle of liquor and the captain consented to a three-hour collecting excursion by some twenty persons among the Subantarctic island's fur seals, king penguins, and rockhopper penguins. "At the very spot where we landed," Wilson wrote, "there lay a big brown seal fast asleep. Out came cameras, hammers, guns, rifles, mauser pistols, clubs, and sketch books, till the poor beast woke up and gazed on us with its saucer-like eyes. . . . The poor beast was sacrificed in the interests of science, as it was a seal we shall not see in the Antarctic. We killed one more and skinned them both." Wilson then proceeded with Scott and Michael Barne to a nearby penguin rookery. "It was rather a butchery, because we wanted some for food as well as for collecting," Wilson reported. "The others had scattered with guns and cameras to try and find anything in the way of eggs or birds or what not, and the result was a really splendid collection for my department." Scott later added, "It was the first time that any of us had seen a penguin rookery, and every detail of their strange habits proved absorbingly interesting; we were lucky enough to have arrived during the nesting season, and were able to collect specimens of eggs and of the young in various stages of development."[8]

After both stops, Wilson devoted days to the exacting task of processing specimens for shipment to London from the next port of call. This mostly involved sketching and skinning birds—the messy chores that then stood at the heart of ornithological fieldwork. "Up at 6 a.m. painting the heads and feet of all my birds. Same all the morning, and then! my word, skinned hard on till nearly 10:30 p.m.," he wrote about the day after the Trindade stop. "They had already begun to stink by the evening." He resumed work the next day, Sunday, following Mass. "It was working under difficulties on deck, where not only the birds you are at, but everything you wanted to use, either rolls away or is blown away. Excellent practice for temper and impatience. And perhaps in that way as good for one as many more prayers." With the Macquarie specimens Wilson received some help. "Skelton, Hodgson and [Jacob] Cross, one of the seamen I have taught to skin, are all hard at work on the penguins," he wrote, "skinning for all they know. I spent the whole day long, from breakfast till the light went, painting the heads and feet of the birds one after another to get the colours before they fade." He painted birds again the next day "till 2 p.m.," then "set to skinning" with the other men until "the light failed us. Spent the evening writing up my notes." Altogether, Wilson concluded, "It will be a pretty big collection."[9]

Following the stopover in New Zealand, collecting resumed as *Discovery* passed through the ice pack guarding the Ross Sea. Here, Wilson obtained his first emperor penguins and Ross seals. Penguins and seals live on the pack during the Antarctic summer and ride it north in search of food. As *Discovery* passed through it, Scott posted marksmen on deck to shoot specimens, with other men dropping onto the floes to secure the booty and haul it on board. "We had not proceeded far into the pack when our upper deck became a busy but gory scene, for in one part men were skinning our prizes in the shape of seals and penguins, whilst elsewhere it was thought advisable to turn our sheep into mutton, and soon we had an array of carcasses," Scott noted. On the day they celebrated Christmas, the explorers saw their first Adélie penguins on the pack. "Peace and goodwill we called them and they were not sacrificed to science," Wilson wrote, but two emperors spied later that day received no such grace. "They proved to be young ones, which I believe have never been procured before," Wilson noted: "A very useful addition to our collection."[10]

The seal and seabird sightings increased again as the ship approached land, especially near the Adélie penguin rookeries at Capes Adare, Jones, and Crozier, which the explorers visited. "Such a sight! There were literally millions of them," Wilson exclaimed about the Adélies at Adare. "In firing at [a] little petrel, although I killed it dead, I also killed five penguins, which gives an idea of how thick they are on the ground, for I only used one [shotgun] barrel." Everyone delighted in watching the penguins; Wilson took copious notes on their habits. "One of the men brought me an albino penguin which has made a good skin and will be very acceptable at . . . the British Museum," he noted at Cape Crozier. "A Giant Petrel and several skuas were caught, but I had to let them go as we have more birds on board already than a month of skinning will finish off."[11] A bay near Cape Jones yielded thirty Weddell seals and ten molting emperor penguins for the ship's larder. "It seems a terrible desecration," Scott wrote, "to come to this quiet spot only to murder its innocent inhabitants and stain the white snow with blood; but necessities are often hideous." Wilson added, "I had to superintend this beastly butcher's work, a duty much against the grain."[12] They did it for science and sustenance.

Research on Adélie penguins promised little new information. Borchgrevink's 1898–1900 expedition had wintered at Cape Adare and studied its penguins. Earlier expeditions had observed these gregarious birds in other Antarctic sites. Given their other priorities, the *Discovery* explorers could not expect to add much to the existing knowledge of Adélies except by collecting more specimens. They brought back twenty-nine skins, eleven adult skeletons, six juveniles in spirits, and eggs from three rookeries.

Wilson's interest turned instead toward emperors, the largest and least known type of modern penguin. At the time, naturalists suspected that they bred on land farther south than any other animal. Before *Discovery*'s voyage, European collections contained only a handful of adult specimens, most dating from Ross's expedition of 1839–43, and one weathered egg from the same period found on an ice floe by French explorers. Wilson wanted to locate an emperor rookery, make extensive collections, and study the birds' behavior.

Early encounters with emperors fed his curiosity. He found only immature ones on the pack, which he depicted as the bird's "nursery or training ground." As the explorers sailed along the Antarctic coast before settling into winter quarters at McMurdo Sound, they came upon a few adults but no large groups until they approached King Edward VII Land at the eastern end of the Ice Barrier. "Here in a bay," Wilson wrote, "we first began to suspect that we were nearing the Emperor's breeding haunts. Away in the distance, over some miles of disintegrating ice-floe, could be seen large companies of birds which, when viewed through the telescope from the crow's nest, proved to be Emperor Penguins." He saw thousands of them. "Obviously these were the Emperors' rookeries, which no one has found before." Wilson wanted to investigate, but it was late in the season and fresh sea-ice was forming around the ship. Scott reversed course and steamed toward winter quarters. "Knowing nothing of their habits, it was natural that we should believe that here at last was the long-looked-for breeding ground," Wilson wrote, "and our disappointment, on realizing that we could not attempt to reach it, may be easily imagined."[13] Later findings would revise his thinking about this colony and help him to better understand the species' peculiar breeding behavior.

At the time, the confused state of evolutionary theory made emperor penguins a fit topic for Wilson's research. Virtually all British biologists, including those on the *Discovery* expedition, accepted the doctrine that current species evolved from preexisting species in a chain of descent extending back to the first appearance of life on earth. In his 1873 account of the oceanographic research that led up to the *Challenger* and *Discovery* expeditions, Wyville Thomson had written, "There is now scarcely a single competent general naturalist who is not prepared to accept some form of the doctrine of evolution." He attributed the general acceptance of this doctrine to "the remarkable ability and candour with which the question has been discussed by Mr. Darwin and Mr. Wallace, and to the genius of Professor Ernst Haeckel," the German evolutionist.[14]

As Thomson implied, however, there was little scientific consensus on the "form" or means of evolution. Biologists still lacked a clear understanding of inheritance or genetics. Even Thomson's three pillars of evolutionism—Darwin, Wallace, and Haeckel—disagreed over how the transmutation process operated. Wallace saw new species generated through the natural selection of beneficial inborn variations in individual organisms. Haeckel favored the more Lamarckian view that selection preserved and propagated adaptive characteristics acquired during an individual's life. Darwin accepted a bit of both views and added a dollop of group selection to account for altruistic behavior. Others proposed inner developmental drives or gross inborn mutations to account for the origin of species. Some of these alternatives so diminished the role of natural selection that, by 1900, some biologists spoke of standing at the "death-bed of Darwinism" in their rush to find non-Darwinian, less selectionist theories of evolution. It almost seemed as if scientists did not want to see evolution driven by a naturalistic struggle for existence among inborn traits. Certainly Wilson did not.

Discovery sailed in the interval after science accepted the theory of evolution and before consensus emerged over how it operated, which followed the rise of population genetics in the 1930s. Wilson hoped that research on emperor penguins might provide insight into the evolutionary process. He understood such research to involve collecting as many specimens as possible at different stages of life in order to study variation within the species, observing behaviors relating to reproduction and survival, and examining embryos for clues to evolutionary history. This final object, which resulted in the explorers going to incredible lengths to obtain penguin eggs, arose from Haeckel's so-called biogenetic law that ontology (the developmental path of each individual) recapitulates phylogeny (the evolutionary history of the group). If evolution proceeded through the accumulation of acquired characteristics added on to earlier types, Haeckel reasoned, then as each subsequent organism developed, it should pass through its ancestral forms before taking on its adult structure. He envisioned this recapitulation of past forms occurring in developing embryos and thought he could use them to trace an organism's evolutionary history much as paleontologists used the fossil record or morphologists used anatomical similarities.

In popular books and academic articles, Haeckel referred to ontogeny, morphology, and paleontology as "the three great ancestral documents" of evolution.[15] Darwin had mostly relied on morphology and paleontology to make his case for evolution in *Origin of Species*, but he took notice as Haeckel claimed to find remarkable recapitulations in the developing embryos of various vertebrate types, including fish, reptiles, birds, and mammals. Later scientists concluded that he exaggerated these similarities, but for a season, comparative embryology served as a powerful argument for evolution. With few fossils expected in the glaciated Antarctic terrain, Wilson would have to look to comparative morphology and embryology to trace the evolution of emperor penguins. Both efforts taxed even his remarkable determination and endurance.

In anticipation of British explorers returning to the Antarctic, Wilson's mentor, P. L. Sclater, had published an article on emperor penguins. In it, he called them "a characteristic form of the little-known Antarctic continent." Naturalists on Cook's 1774–75 voyage brought back the first reports of these large swimming birds, but no specimens. At first, European ornithologists considered them a larger variety of the better-known and similar-looking king penguin. Ross and the American explorer David Wilkes collected the first specimens of emperors in the 1840s, leading to their classification as a separate species within the same genus as the king—the only two penguins in this grouping—though Sclater found the skeletons of the two types different enough to justify splitting them into distinct genera. "As regards the exact range of the Emperor Penguin," Sclater wrote, "much remains to be learned, but it appears to be now only found on the shores of the Antarctic continent, and probably breeds in the adjacent islands." Based on Ross's account, he noted how hard they were to catch and kill.[16]

Wilson thought the emperor was the oldest surviving species of penguin, presumably because of fossils suggesting that even larger penguins had lived in the past. Further, because he viewed their feathers as more akin to fish scales than the feathers of other birds, Wilson described penguins in general as "the most primitive, behindhand birds in existence."[17] He saw the emperor as a critical species for unpacking the mysteries of avian evolution and likely believed that God had graced him with the opportunity of studying them first.

After settling into winter quarters, the explorers did not see another emperor for almost two months. On Easter Sunday, March 30, a large party of penguins surprised the explorers by wandering from south to north past the icebound ship. Wilson reasoned that the birds must be leaving their summer rookery in the south for open water in the north. Preoccupied with preparations for a group sledging trip the next day, Wilson did not capture any of these birds but made ready for the next encounter. It came on April 8, and led to what he later called the "Great Emperor Penguin Hunt."[18]

"Just as I sat down to dinner, news was brought that some 30

Emperors were in sight," Wilson wrote in his diary. Donning a thin balaclava, he rushed out with Barne, Cross, and Petty Officer Thomas Kennar into the Antarctic night. "It was blowing a blizzard at the time with drifting snow and very cold," he observed. "Wul, us sets out," Kennar reported with a mock-Cockney accent in the expedition's newspaper, "foure warriors bolde." They caught up with the penguins about a mile from the ship and drove them back toward it. Others brought out wire netting and joined in trying to corral the birds. "We couldn't get the penguins to remain in the wire netting," Wilson wrote. "Three times they broke through and scattered and we rounded them up again, till at last we thought we had better kill what we wanted." Kennar elaborated, "Talk about blood upon a battle field!! I wus a'hangin on to their heads while Chucks 'e was a'sawin um orf. Then us got roun' um again, and they charged us agin! . . . Then us all sets too an the fearful slighter begun. Knives wus snapped off close to the hilt. Knives was lost an I has to get out I'se pusser's dirk." So it went until most penguins in the huddle were dead. "It was not a pleasant job," Wilson conceded, "but an opportunity not to be missed, for no one has ever seen this bird in any numbers before, and it is our duty to bring back as good a collection as we can."[19]

Two-thirds of the adult penguin skins in *Discovery*'s collection came from this encounter, including one from the largest bird obtained, which weighed in at ninety pounds. Alone or in small groups, fifteen more emperors passed the ship in April and May, before winter set in, and then no more until November. By then, Wilson had a new focus for his emperor penguin studies.

After the disastrous March effort to post notice at Cape Crozier of *Discovery*'s wintering site, Scott sent Charles Royds, Reginald Skelton, and four others back in October to try again. Successful this time following the fifty-mile trek, Skelton took the opportunity to photograph the region and, while doing so, spied hundreds of emperors offshore on the sea ice. "He asked himself what they could be doing here in such numbers," Scott wrote, "and wondered if it were possible that at last the breeding-place of these mysterious birds had been discovered —it seemed almost too good to be true."[20] Delayed in his effort to reach the site by a howling blizzard that kept the party in their tent without access to a stove or water for five days, Skelton finally climbed down to the sea ice on October 18, making his way over a series of pressure ridges and into the rookery. There, he found more than four hundred adults, including some still nursing chicks, but no eggs. The party returned to the ship with three freshly killed chicks, a small box of broken egg shells, photographs, and stories of survival in temperatures below −50°F. The explorers could only marvel that emperors raised chicks in such conditions.

From Skelton's discovery, Wilson began piecing together an entirely new account of how emperors breed. They must lay and hatch their eggs in early spring on the sea ice rather than in summer on land like other penguins. He reasoned that this would allow the chicks to reach sufficient size to protect themselves from avian predators such as skuas and petrels before those migratory birds returned to the region. Wilson had seen what havoc such predators wrought at Adelie rookeries and could see how early breeding might help emperors. Skelton's pictures showed each chick balanced on the feet of an adult under a protective flap of the adult's skin, much as happened with king penguins. The eggs must nestle here too, Wilson deduced, insulated from the sea ice during incubation.

The chicks interested Wilson as much as the adults. "They are very handsome too and of a colour I had not suspected," he wrote. "The back is covered with a very soft white fluffy down which deepened into gray on the under parts. The forehead, head and neck were jet black." No one had ever seen emperor chicks. Their color, which Wilson called "a fantastic piebald of black and white," differed markedly from the uniform darkness of king penguin chicks. Yet adults of the two species looked alike. He could not understand how such a color difference could evolve in closely related birds unless it served a distinct purpose. Wilson dismissed the obvious explanation that the white down served as protective camouflage, because the chicks had no predators and their black heads still stood out against the sea ice. "The white down of the chick must be a special development of his own, probably upon the lines of physiological economy," he concluded. "Such pigment as it was able to produce instead of being squandered over the whole body, as it is in the young King Penguin, is in the young Emperor concentrated in the head as black, where it may be used as a signpost to its parent in the pack." Wilson admitted that he was "falling back on what is still an uncertain theory," but the effort showed how sincerely he rejected the role of a mere collector and sought to participate fully in the scientific enterprise.[21]

When the Crozier party returned with its report, Wilson wanted to visit the rookery. But Scott had chosen him for the Southern Sledge Journey, on which Markham and Scott pinned their quixotic hopes for the pole, and its impending departure did not leave time for an excursion to the cape. "I am afraid this long southern journey is taking me right away from my proper sphere of work," Wilson complained in his diary, "to monotonous hard work on an icy desert for three months, where we shall see neither beast nor bird nor life of any sort nor land and nothing whatever to sketch."[22] Instead, Scott authorized Royds to lead a small party back to the rookery in search of fresh eggs that might contain embryos. This party departed on November 2, the same day that Scott's team headed south, so Wilson had to wait until his return in February to hear Royds's report. It raised more questions than it answered.

Royds reached the rookery on November 8 to find the chicks and much of the sea ice gone. Only some adults remained. Within two days, these had left too. "The migration to the North was obviously complete, but how had the chicks been taken?" Wilson asked. "One thing was amply certain, that before their down was shed they could not have gone by water." If the expedition stayed for a second spring, Wilson resolved to visit the rookery himself. "It appears then that in the Emperor Penguin we have a bird which not only cannot fly," he speculated, "but which never steps on land or on land ice even to breed, and has so modified its habits that it carries out the whole process of incubation on the sea ice, choosing those months of the Antarctic year when the greatest cold ensures a solidity of sea-ice which can be trusted."[23]

Scott shared Wilson's fascination with emperors. "All other birds fled north when the severity of winter descended," he commented on Royds's trek, "and we gathered no small satisfaction from being the first to throw light on [its] habits."[24] Royds's party did find one abandoned emperor egg half-buried in the snow, but its putrid conditions offered few clues to the living embryo it had once contained.

When *Discovery* remained icebound for a second winter, Wilson began hatching his plans. "I have been ferreting out various things in all the books I can lay hand on, to try to get an idea of how long the Emperor Penguins ought to sit on his eggs before hatching out," he wrote during the depths of that second winter. "I have come to the conclusion that the eggs must be laid in the middle of August. They are incubated then for at least seven weeks, i.e. all through September, the coldest month of the year, and the chicks are taken up northward on floating ice at the end of October."[25] He would go to Cape Crozier in early September to collect eggs and again in mid-October to watch the chicks depart. If he left much earlier, the first trip would take place during the winter darkness. Again, however, the penguins upset Wilson's plans.

Royds led the September trek, which included Wilson, Cross, and three others. "It was nasty weather," Wilson wrote. The temperature dipped below −60°F, lower than on any previous outing, and on some days never rose above −50°F. With limited daylight for sledging, it took six days before the explorers surmounted the final pressure ridges to enter the rookery, where they found a thousand adults and up to two hundred chicks. "What then was our dismay when we found that we were again far too late for eggs, and that every one of them was hatched," Wilson lamented. He concluded that emperors must lay their eggs in July and hatch their young in August![26]

Arriving in September, the explorers could merely observe nursing behavior, collect abandoned eggs, and gather some chicks, two of which Cross carried back alive. Wilson hoped to raise the chicks at least until they molted, but both died within three months. This was sufficient time, however, to support Wilson's hypothesis that wild chicks migrated north on sea ice before they molted and could swim. At the rookery, moreover, the instinct of some adults to fight over and brood abandoned eggs and dead chicks, whether or not their own, so impressed Wilson that he thought that they must pass around fresh eggs and live chicks too. "Promiscuity seems to be a marked feature and I think each chick must change hands scores of times," he commented in his diary.[27]

These findings provided little consolation for Wilson, who wanted most of all to explain avian evolution. "The possibility that we have in the Emperor Penguin the nearest approach to a primitive form not only of a penguin, but of a bird, makes the future working out of its embryology a matter of the greatest importance," he observed. "It was a great disappointment to us that although we discovered their breeding ground, and although we were able to bring home a number of deserted eggs and chicks, we were not able to procure a series of early embryos by which alone the points of particular interest can be worked out." He now knew that this would involve a midwinter trek to the rookery in nighttime darkness during a later expedition. After giving the venture some thought following his return to Britain, the indefatigable Wilson concluded, "The whole work no doubt would

E.A. Wilson, del.

EMPERORS, BARRIER AND SEA-ICE

Edward Wilson's sketch of the emperor penguin rookery on the sea ice under the Great Ice Barrier at Cape Crozier, from Apsley Cherry-Garrard's *Worst Journey in the World* (New York, 1922).

be full of difficulty, but it would not be quite impossible." It might be the worst journey in the world, but it was not beyond him to try it should he get back to Antarctica.[28]

The October trip to the rookery proved more successful. As planned, it coincided with the emperors' migration north. Wilson led this trip, which included Cross and Royal Navy Stoker Thomas Whitfield, a giant of a man. To watch the migration, they camped near a cliff above the sea ice for three weeks, subsisting mostly on fresh seal and penguin meat cooked over an improvised blubber stove. "Stewed, they were delicious," Wilson wrote of the penguins, "but fried in butter in 'blessed mouthfuls' they were heavenly." Little had changed since the September trip except that more chicks had died and the survivors had grown larger. "The largest chicks were neatly all tucked in under the parent's flap, head first, so that only a large ball of gray fluff with a stumpy black tail appeared. At times one saw the chick's head poking out behind," Wilson noted. "Old birds could be seen here and there nursing dead chicks, a most pathetic sight, showing how very strong is the desire to brood."[29]

Within days after the party arrived, a gale started blowing from the south. In an apparent response to the weather, the emperors began streaming in long lines from their sheltered rookery to the edge of the sea ice. "Large sheets were going out leaving open sea," Wilson observed on October 23. "Some 200 or 300 emperor penguins were waiting on the thin ice edge and others were filing out to join them." Some walked freely; others waddled with chicks below. The worsening weather kept the explorers in their tent off-and-on for ten days, but each time they could look, more sea ice and penguins were gone and a new group of penguins stood at the ice's retreating front. "Some 300 were waiting there today," Wilson wrote on October 25. He counted only 400 adults and 30 chicks at the rookery on November 2, when his party departed for the ship, and 63 dead chicks. Added to the dead chicks and deserted eggs gathered in September and compared with the number of live chicks observed on both trips, Wilson estimated that only about one in four emperors reached maturity. This death rate, he concluded, coupled with the strong instinct to brood led to what he depicted as a mad rush by lone adults to claim unprotected eggs and chicks. With this, Wilson had seen enough to propound his account of the emperor penguin's life cycle.[30]

Upon his return to Britain following the *Discovery* voyage, Wilson agreed to write the chapters on birds and sea mammals for the expedition's official reports. These reports, which appeared in ten folio volumes published by the Royal Society or British Museum from 1907 to 1912, represented the expedition's principal scientific contribution.

"I want my monograph on the Emperor Penguin to be a classic," Wilson wrote to his father.[31] It related his experiences with emperors, featured his observations of their behavior, and incorporated his theories about their life cycle. In all, it provided a compelling new narrative about an engaging species. Wilson depicted emperors living at the sea in the summer, marching south to rookeries in the fall as other Antarctic birds migrated north, laying and hatching eggs in winter, raising chicks in early spring, and riding north on the sea ice in late spring.

Turning to their social evolution, Wilson presented emperors as a textbook case of group selection. "Incubation is carried out not by one bird only, nor by a single pair, but by a dozen or more," he wrote. "Hunger induces a bird to pass on its chicken to someone else while it goes to feed and on returning is perhaps content to join the ranks of the unoccupied until hunger induces the desertion of another chick." He then added, "Why this should be the case is more than one can say; possibly it is a condition of things evolved in an exacting climate, to allow each adult to obtain sufficient food through so long a period of incubation."[32]

This view of emperors reflected Wilson's own Antarctic experience. Survival required sharing the work, he learned, and he did far more than his share. He "helps with any job that may be at hand," Markham noted, "and in fact is an excellent fellow all round."[33] Later research on emperors found that the male alone incubates the egg, losing over half its weight in the process, until the female returns with food for the chick. No others help. Wilson missed this part of the story. "As there is no such thing to be found in September as a bird half-starved," he wrote, "it is obvious that the same bird does not sit on the same egg for seven weeks."[34] He believed it took a group to survive in the Antarctic, and his theistic view of evolution left room for nature to endow altruism in even the most primitive of birds. The scramble for chicks and eggs that he interpreted as a group instinct to brood unprotected offspring was subsequently seen as the ineffective acts of inexperienced breeders rather than a species trait.

Back in Britain, Wilson took up the cause of protecting penguins from harvest for their oil. He lectured on birds, conducted a government-funded study of grouse disease, and illustrated books on birds and mammals. When Shackleton mounted his *Nimrod* expedition in 1907, he asked Wilson to serve as his second-in-command. At the time, however, Wilson felt obligated to finish his grouse study. Further, after Shackleton announced his intentions, Scott told Wilson of his own plans for a second expedition and invited him to serve as the leader of its scientific staff. Knowing he would have time to complete the grouse study first, Wilson agreed. In the meantime, to extend *Discovery*'s scientific work, Shackleton would need to assemble a fresh team of researchers.

When announcing his expedition, Shackleton assured the RGS leadership that "I shall in no way neglect to continue the biological, meteorological, geological, and magnetic work of the *Discovery*."[35] He wanted his mission to best Scott's in every respect, including science, and it went a long way toward doing so, often in spectacular fashion. In magnetic work, for example, Shackleton's team actually reached the South Magnetic Pole while Scott's team fell short. In meteorology, both expeditions had used Mount Erebus as its weather vane and Scott's men had talked about climbing it, but only Shackleton's did so. In oceanography, the *Nimrod* expedition continued winter dredging and netting at least on a par with *Discovery*'s efforts. Of course, in terms of geographical discovery and human endurance, Shackleton's Southern Sledge Journey far surpassed Scott's Farthest South. In each case, the *Nimrod* expedition clearly continued and extended *Discovery*'s research. Similar parallels would appear in geology and glaciology. The pattern broke only in the land-based biological work of the two expeditions. Even though Shackleton established winter quarters on the site of an Adelie penguin rookery at Cape Royds, his biologist, James Murray, did not attempt to extend Wilson's research on Antarctic birds.

In his monograph, which Murray could have reviewed before departing for Antarctica, Wilson described how someone could extend his work on emperor penguins by collecting and dissecting embryos and young chicks in midwinter at Cape Crozier. He even detailed how and when such a trip should take place. By climbing Mount Erebus and sledging toward both poles, Shackleton and his men showed a readiness to accept challenges, but they ignored this one. Perhaps they thought it impossible to sledge sixty miles over rough terrain during the depths of the cold, dark Antarctic winter. In Murray, further, Shackleton had a biologist more interested in freshwater microorganisms than birds, and having recently suffered a physical breakdown, Murray rarely ventured far from base. The expedition's second oldest member at age forty-two, he had taken up science in midlife and worked under *Challenger's* John Murray (to whom he was not related) on a study of Scottish lakes. If Shackleton had wanted to build on Wilson's research, he would have chosen an ornithologist instead of a limnologist. Of course, Shackleton tried to recruit Wilson for the expedition, and by the time he settled on Murray, he knew that Wilson, whom he greatly admired, would return to the Antarctic with Scott. Shackleton may have wanted to leave the penguins to him. In Murray, the *Nimrod* expedition gained a gifted biologist who took his research in an entirely different direction from Wilson.

Murray's field notes depicted Cape Royds as a two-hundred-acre promontory of rocky hills, gravel-filled valleys, and freshwater lakes jutting into McMurdo Sound at the base of Mount Erebus. Some exposed ridges and moraines extended for a few miles behind it. "Beyond the limits of the small area thus briefly described," Murray noted, "snowfield and glacier stretched for many miles, offering no support for any living thing, unless it be some of those lowliest organisms which can exist on the surface of the snow itself."[36] This small tract of exposed land and lakes became his outdoor laboratory—and he found life there beyond anyone's wildest expectations.

Discoveries came slowly. "It is difficult to imagine a more unpromising field for biological study," Murray observed, "than Cape Royds appears on first examination. Nothing is to be seen but a succession of ridges of black lava (shattered into loose blocks by the intense cold), moraines and snowdrifts, all apparently equally barren." No flowering or higher plants appeared anywhere on land. Here and there on the moraines above the cape, Murray and his chief assistant, the twenty-one-year-old English geology student Raymond Priestley, did find some tufts of moss and bits of lichen, but an expert who later examined the collected specimens described them as "sickly plants, struggling painfully against exceptionally hard conditions."[37] Priestley and other members of the expedition also collected mosses and lichens at a few other Antarctic sites and brought them back to Murray, but he characterized all these specimens as decrepit. In the Northern Hemisphere, he noted, a far richer flora existed at higher latitudes.

As a microbiologist who had studied Scotland's cold-water lakes, however, Murray knew where to look in such unpromising terrain, and he put that knowledge to good use. First he found bright green algae on a small frozen lake near the hut. Then others spied brilliant orange algae in the transparent ice of one of Cape Royds's larger lakes. "Later on, this plant was found to be abundant in all the lakes," Murray noted, "and in many of them it formed continuous sheets over the entire bed." Ultimately, he identified eighty-four species of algae in the region, from three distinct classes. Priestley went to work chipping out samples and in late winter dug a shaft through fifteen feet of solid ice to reach the bottom of the largest lake. Some of the larger lakes never thawed during the expedition's thirteen months at Cape Royds. Other, smaller ones got as warm as 60°F in summer. All of the cape's lakes and ponds contained algae, and frozen or unfrozen, algae

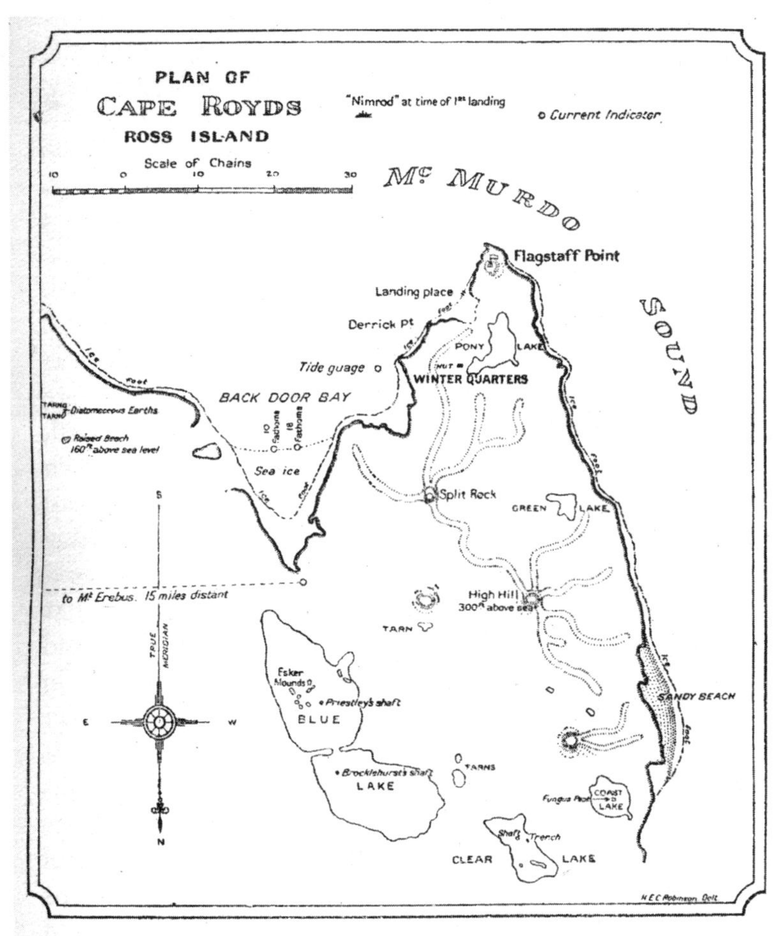

WINTER QUARTERS

Map of the freshwater lakes around the *Nimrod* expedition's winter quarters, from Ernest Shackelton's *Heart of the Antarctic* (Philadelphia, 1909).

from all of them supported minuscule animals—mostly wormlike rotifers and tiny water bears, but also nematode worms, mites, and small crustaceans. "Everywhere, microscopic life swarmed on this weed," Murray declared, with some of the species new to science.[38]

Once he discovered them, Murray set about studying these small animals. In summer, if a lake thawed, they thrived, and in winter they survived in a dormant state. As soon as Murray thawed a frozen piece of algae, the animals on it would spring to life. "To test the degree of cold which they could stand," he wrote, "blocks of ice were cut from the lakes and exposed to the air in the coldest weather of the whole winter. By boring into the center of the blocks we found that they were as cold as the air. A temperature of minus 40° Fahr. did not kill the animals. Then they were alternatively frozen and thawed weekly for a long period, and took no harm. . . . At last they were dried, and the bottle containing them was immersed in boiling water, which was allowed to cool gradually and still a great many survived." Even salt water did not kill them. "It is not the eggs merely that survive all these changes, but the grown animals," Murray marveled. "They may be frozen solid, their very blood frozen, and they may remain thus for many years, but always alive and ready at a moment's notice to resume feeding whenever the ice melts."[39] He tried to collect and identify as many types as possible. Nothing in their behavior necessarily set them apart from similar microorganisms found elsewhere, but their sheer existence in these lakes extended the known range of life. British naturalist and polar explorer R. N. Rudmose Brown later credited Murray with launching "a new field of Antarctic research."[40]

Murray's enthusiasm for microorganisms was contagious. It drew in Priestley, who persevered in extracting specimens from Cape Royds's frozen lakes while Murray was indisposed by such severe chronic diarrhea that his cubicle in the expedition's hut became known as the Taproom. "Without him the greater part of the collections would not have been made," Murray said of Priestley. "It does not lessen our indebtedness to him to tell that he enjoyed his self-imposed task, and his voice might be heard issuing in light-hearted song from some deep shaft in lake or sea." Others helped as well. Edgeworth David collected samples of algae during his Northern Sledge Journey to the magnetic pole, Douglas Mawson contributed some from high moraines on Mount Erebus, and even Shackleton retrieved algae from a lake near Hut Point. In his summary of the expedition's accomplishments, Shackleton later hailed the discovery of microscopic life in Antarctic lakes as "the most important event of the winter months."[41] Upon his return, Murray filled a large volume with his findings.

Even as Murray conducted his research and assembled his findings, Wilson plotted his return to Cape Crozier's emperor penguin rookery for fresh embryos. Such a journey from winter quarters on McMurdo Sound, he wrote in his 1907 monograph on penguins, would "require that a party of three at least, with full camp equipment, should traverse about a hundred miles of the Barrier surface in the dark and should, by moonlight, cross over with rope and axe the immense pressure ridges which form a chaos of crevasses at Cape Crozier." The cape sat at a convergence of glacial forces: barrier ice turned east by Ross Island met barrier ice moving north, causing massive disruptions in the surface. Further, Wilson noted, due to its location at the base of Mount Terror, storms from the south bear down on the cape with awesome fury. No one had ever tried sledging in the Antarctic winter, much less to as remote a location as Cape Crozier, yet Wilson envisioned more than a mere sledge journey. "The party would have to be on the scene at any rate early in July," he stressed, and "it would by this time be useful to have a shelter built of snow blocks on the sea ice in which to work with the cooking lamp to prevent the freezing of the egg before the embryo was cut out, and in order that fluid solutions might be handy for the various stages of its preparation." As they would be deformed by freezing, Wilson believed that the embryos must be processed on-site.[42]

Wilson started making arrangements for the so-called Winter Journey soon after he agreed to join Scott's second expedition, and he continued working out details until the journey began. He wanted to reach the rookery during the first winter and proposed wintering the shore party at Cape Crozier to facilitate his research. Scott opted for a base on McMurdo Sound and tried to talk Wilson out of the Crozier trip but ultimately let him go in spite of the risks. Wilson also discussed the trip in advance with Apsley Cherry-Garrard, a recent Oxford graduate whose £1,000 contribution toward the expedition's costs secured him a place as the assistant biologist despite his extreme nearsightedness. To complete the Crozier party, Wilson chose Birdie Bowers, a short, stout marine officer who signed on as the ship's storekeeper but won a place on the shore party with his boundless energy and organizational ability. "This winter travel is a new and bold venture, but the right men have gone to attempt it," Scott wrote on the day they departed.[43] They packed a double-lined tent, a blubber stove, and eiderdown liners for their reindeer sleeping bags—all improvised in the weeks leading up to the trip. To gain insight into human physiology from their extreme effort, the men took only biscuits, butter, and pemmican for food and experimented eating different amounts to gauge the optimal proportion of carbohydrates and fat for sledging. Wilson expected to be out for six weeks.

The Winter Journey exemplified the extreme lengths to which the *Terra Nova* expedition went for science. "Wilson came to the conclusion that Emperor embryos would throw new light on the origin and history of birds, and decided that if he again found his way to the Antarctic he would make a supreme effort to visit an Emperor rookery during the breeding season," the scientist who examined his findings explained. Expedition geologist Griffith Taylor added, "Wilson hoped to get embryo chicks, and thus study the early stages of these birds, which in some way are the most primitive existing, and which therefore exhibit features linking them to the reptiles." In particular, Wilson wanted to see if the youngest embryos of this supposedly oldest avian species displayed vestiges of reptilian teeth and whether the papillae that became their nesting feathers corresponded to those that develop into reptiles' scales. If ontology recapitulated phylogeny, as Haeckel said, and Wilson could find evidence in penguin embryos linking two classes of animals, then he would confirm that birds descended from reptiles and reinforce the case for evolution. "One hopes to find real teeth in the embryo of the emperor penguin, though none are present in the adult bird," Wilson told his colleagues shortly before departing for Cape Crozier.[44]

The Crozier party left winter quarters at Cape Evans on the new moon—June 27, 1911—five days after the winter solstice. With neither sun nor moon, it was the darkest day of the year, but leaving then promised to put the party in the most heavily crevassed part of its outbound journey during the full moon and get it to the rookery early in the nesting season. Roughly speaking, Ross Island forms an equilateral triangle with sides about fifty miles long. Cape Crozier lies at its eastern tip, Hut Point and Cape Armitage at its southern end, and Cape Bird at its northern apex. Cape Evans juts from the island's west coast about fifteen miles north of Cape Armitage. Starting there and following the coastline around to Cape Crozier made the journey more than twelve miles longer than it was from Hut Point on the *Discovery* expedition, for a total of nearly seventy miles each way.

With scientific equipment, cold-weather gear, and supplies loaded onto two nine-foot-sledges linked end to end and to the men by harnesses, the party man-hauled more than 750 pounds. "The three travellers found they could pull their load fairly easily on the sea ice," Scott noted when they left. "I'm afraid they will find much more difficulty on the Barrier." Taylor added, "No one realized what they would have to encounter, and I hope no one will ever again attempt to do anything so close to the confines of human endurance." He joined a small group that accompanied them for the first three miles. "When we stopped I called for three cheers for the Cherry Winter Knight [Cherry-Garrard], the Short Winter Knight [Bowers], and the Long Winter Knight [Wilson]. When they saw that I meant 'Knight' (and not the surrounding gloom!) they laughed muchly."[45]

Wilson was as resolute as Scott, and together they set the expedition's tone. One day after the Crozier team left, Scott opened a fresh notebook by quoting T. H. Huxley: "The highest object that human beings can set before themselves is not the pursuit of any such chimera as the annihilation of the unknown; but it is simply the unwearied endeavour to remove its boundaries a little further." A few days later, he added his own aphorism: "The fact is science cannot be served by 'dilettante' methods, but demands a mind spurred by ambition or the satisfaction of ideals." These words reflected both Scott's view of Wilson's Winter Journey and a perspective on science, exploration, and life that these two comrades shared.[46]

The outbound journey went as well as one could expect. The men traveled from Cape Evans south to Cape Armitage and from there onto the Ice Barrier. Then it got cold—colder than ever previously recorded in Antarctica—as they traced the icebound side of Ross Island from Cape Armitage to Cape Crozier. "The min. temp. last night was −75.8°," Wilson wrote on July 7. "At 2 p.m. −58.3° and at 7 p.m. −55.4° a rise for which we were grateful as it is distinctly easier to save one's feet from frost bite at −50 than at −70." The heavy surface caused by the extreme cold forced the men to relay the sledges—one mile gained for three miles walked—so that they typically netted only about two miles per day during this stage of the journey. "Our feet often went through the soft snow and slipped on the hard wind-swept sastrugi underneath," Wilson added on July 9. It was like trying to pull heavy sleds through a sandy desert in total darkness at −60°F. Everything froze. "The trouble is sweat," Cherry-Garrard explained. "It passed just away from our flesh and then became ice. . . . When we got into our sleeping-bags, if we were fortunate, we became warm enough during the night to thaw this ice: part remained in our clothes, part passed into the skins of our sleeping-bags, and soon both were sheets of armour-plate."[47]

The temperature rose into the minus twenties Fahrenheit as the men neared the Ice Barrier's end at Cape Crozier. The surface became firmer with the warmer weather, permitting them to pull both sledges at once. Yet now, with the broad Ice Barrier mostly behind them, they needed to zigzag by the faint light of the waxing moon between the crevassed land ice on their left and the worsening pressure ridges on their right. "It was most difficult to keep the middle way between the traps," Wilson wrote on July 14. "In daylight, in the *Discovery* days, we used to have but little difficulty here, but we were never here in the dark." They took turns falling into crevasses and bumping into hummocks of ice. "Crevasses in the dark *do* put your nerves on edge," Cherry-Garrard observed—but he could barely see anyway because it was too cold and windy to wear glasses while sledging.[48]

Finally, on July 15, nearly three weeks after they began, the men climbed onto a windswept moraine overlooking Cape Crozier. "The view . . . was magnificent and I got my spectacles out and cleared the ice away time after time to look," Cherry-Garrard wrote. "To the east a great field of pressure ridges below, looking in the moonlight as if giants had been ploughing with ploughs which made furrows fifty or sixty feet deep. . . . Behind us Mount Terror on which we stood, and over all the gray limitless Barrier seemed to cast a spell of cold immensity. . . . God! What a place!"[49] Somewhere on the sea ice about four miles east, shrouded by darkness, lay the rookery.

Before trying to reach it, Wilson, Bowers, and Cherry-Garrard built a stone hut on the moraine, where they planned to live and work while studying the emperors. It took three days of hard labor to complete the project. Using a sledge as a ceiling beam, they stretched canvas over the top of four rock walls, caulked cracks with snow, and assembled the blubber stove for heat. The tent stood just outside. To protect it from storms sweeping off the Ice Barrier, they built the hut into the lee side of the moraine, just below the ridge. "It seems too good to be true," Cherry-Garrard wrote about settling in at Cape Crozier. "Surely seldom has any one been so wet—our bags hardly possible to get into, our wind-clothes just frozen boxes. Birdie's patent balaclava is like iron—it is wonderful how our cares have vanished."[50] With the hut up and in need of blubber for fuel, the party set out for the penguins.

Using the midday twilight that now dimly illuminated the ice for a few hours each day, the men tried twice to reach the rookery. The first time, they dropped onto the Ice Barrier beyond the first pressure ridge and spent five hours wandering amid ice hummocks before retreating to the hut at dusk. "I went into various crevasses at least six times," Cherry-Garrard reported of their ordeal in this maze of ice. "To be caught in the night there was a horrible idea." Wilson added, "We had the mortification of hearing the distant cries of some of the Emperor Penguins echoed to us from the rock cliffs," but they could not reach them. The next day, they made a steep descent to a narrow gorge between the first pressure ridge and the land-ice cliff, which they followed toward the rookery. They encountered obstacles where ice had pushed in from offshore or fallen from the cliff, but they either crawled through holes under them or cut steps in the ice to climb over them. "At last we got out on to the ice foot overlooking the sea ice, and there were the Emperor Penguins," Wilson wrote. "The light was rapidly failing when we at last reached the sea ice, and we had to be very quick in doing what we had to do here," Cherry-Garrard noted.[51]

The explorers faced a dilemma. "After indescribable effort and hardship we were witnessing a marvel of the natural world," Cherry-Garrard wrote. "We had within our grasp material which might prove of the utmost importance to science; we were turning theories into facts with every observation we made,—and we had but a moment to give." The party had not carried supplies to stay out, nor could they get back in total darkness. "If we got benighted here we should have a nice long time to put in in the dark without food or shelter," Wilson wrote. So they quickly skinned three penguins for blubber, grabbed six eggs, and bolted for the hut with the expectation of returning later for more research. After repeatedly getting lost, struggling up cliffs, and stepping into ice cracks, the party reached the hut with three intact eggs but in no position to examine the embryos. "As we groped our way back that night," Cherry-Garrard observed, "sleepless, icy, and dog-tired in the dark and the wind and the drift, [death by falling in] a crevasse seemed almost a friendly gift."[52]

The weather deteriorated overnight and threatened to become much worse. By situating the hut just below the ridge, Wilson hoped to protect it from wind, but he had placed it in the path of powerful updrafts. The storm's first strong gusts ballooned the canvas roof upward and sucked snow in through every crack. The men piled slabs of ice on the roof during a lull in the storm, but it made little difference.

The brunt of the storm hit the following night with gale-force winds that carried away the tent. "The roar of the wind in the [hut] sounded just like the rush of an express train through a tunnel,"

Cherry-Garrard recalled. "We could only talk in shouts." The hut's roof rose and fell with growing violence until finally, about noon on the third day, it exploded outward in shreds, leaving the men cowering in their sleeping bags under drifting snow and rubble from the roof and walls. "There was no choice for us now," Wilson wrote, "we had to remain lying there in our bags till the blizzard stopped." It was his thirty-ninth birthday. Seventy miles from winter quarters without a tent or shelter, Cherry-Garrard lost all hope and began thinking about taking morphine to ease the end. The others never lost faith. "I was resolved to keep warm," Bowers wrote in his diary, "and beneath my debris covering I paddled my feet and sang all the songs and hymns I knew to pass the time." Although their religious beliefs differed—Bowers being an idiosyncratic proto–Jehovah's Witness—Wilson joined in the hymns while weighing the options. "We could build a snow hut each night on the way home," he noted in his diary, "or we could always dig a burrow in the Barrier big enough for the 3 of us. . . . We had no doubts about getting back so long as this blizzard didn't last till we were all stiffened with the cold in our bags."[53]

The storm subsided a day later, and Bowers found the tent neatly deposited on a snow slope about a quarter mile away. He wanted to try again for the rookery, but Wilson overruled him. "It was disappointing to have come all this way and to have done nothing worth mentioning of our work with the Emperor Penguins, but to remain now," Wilson concluded, "had become a practical impossibility. We had to own ourselves defeated by the Cape Crozier weather and by the darkness."[54] Wilson had promised Scott he would bring the party back alive, and his words governed his actions. In the hope of returning someday to resume his penguin studies, he left behind pickling solution and other research apparatus. He viewed the dire situation as merely a temporary setback.

Even with a tent, the return trip was a living hell. The exhausted party first had to recross the Scylla and Charybdis of pressure ridges and crevassed land ice around Cape Crozier, then pass over the frigid Ice Barrier to Cape Armitage, before turning north on the sea ice to Cape Evans. With only four hours of dim twilight and up to sixteen hours of sledging each day, darkness remained a severe handicap. "We travelled as much by ear and by the feel of the snow and ice under foot as by sight," Wilson wrote.[55] At one point, Bowers dropped into a deep crevasse and hung by his sledging harness out of sight until rescued with a rope.

The temperature again plunged toward −70°F, making the sleeping bags worthless for warmth. "I still kept my down lining in my bag though it was flat as sheet tin and about as warm and soft, but it held my reindeer skin bag together," Wilson noted. So that they could get in them the next night, the men let their bags freeze open each morning and laid them flat on the sledge. By the end of the trip, each seventeen-pound bag had accumulated up to twenty-seven pounds of ice. Cherry-Garrard complained of uncontrollable shivering and recurrent nightmares whenever he tried to sleep. "The worst job," he wrote, "was to get into our bags: the second or equal worst was to lie in them for six hours." None of them got much sleep at night; Bowers and Cherry-Garrard became so exhausted that they literally fell asleep while sledging. Cherry-Garrard's jaws chattered so much that all of his teeth shattered. "The journey home," Wilson concluded, "was by far the coldest experience I have ever had," and he had been on Scott's brutal Southern Sledge Journey of 1902–3.[56]

In his book about the expedition, Cherry-Garrard described the Crozier trip as "the worst journey in the world." "No words could express its horror," he wrote. Still, he noted, "We kept our tempers—even with God." Wilson viewed the ordeal as some sort of divine blessing and believed that God watched over the Crozier party every step of the way. The return trip took eight days, and the entire journey lasted just over five weeks. Wilson told Cherry-Garrard that "it was infinitely worse than the Southern Journey in 1902–3."[57]

Other members of the expedition gasped when they saw how the Crozier party appeared upon its return and shuddered when they heard the travelers' tale. "Cherry staggered in looking like nothing human," Taylor wrote. Surveying all three men, Scott elaborated, "They looked more weather-worn than anyone I have ever seen. Their faces were scarred and wrinkled, their eyes dull, their hands whitened and creased from the constant exposure to damp and cold."[58] After hearing their account, Scott comforted them by saying, "You know, this is the hardest journey that has ever been made."[59] The expedition's newspaper celebrated Wilson, whom the explorers called Uncle Bill, with a poem stressing the mission's scientific purpose:

> YOU were bold, Uncle William, that journey to make,
> Setting blizzards and cold at defiance,
> Cape Crozier to seek, the hut to forsake,
> You're a regular martyr to Science.[60]

In his personal journal, Scott suggested that the Winter Journey had greater emotive than scientific value. "Wilson is disappointed at seeing so little of the penguins," he wrote, "but to me and to everyone who has remained here the result of this effort is the appeal it makes to our imagination as one of the most gallant stories in Polar History. . . . It makes a tale for our generation which I hope may not be lost in the telling."[61]

Some biologists with the *Terra Nova* expedition pursued other types of research during the first winter. After the sea ice formed, for example, Edward Nelson tow-netted for marine specimens in the manner of *Discovery*'s Thomas Hodgson and *Nimrod*'s James Murray at a station off Cape Evans less than a mile from winter quarters. He once visited lakes around Cape Royds for freshwater specimens. Remembering Hodgson's labors and knowing what Murray had done, however, Scott expressed "disappointment" with Nelson. "A clever fellow," he wrote, "but idle." In his journal, Scott added, "Here are a fine opportunity and a not inconsiderably outlay wasted on a young man whose habit in life is that of a pot house politician." It frustrated Scott that a scientist would waste a chance to conduct research through idleness. "I scarce know how to deal with the case," he wrote.[62] Charles Wright noted Nelson's "taste for gin and bridge"; Taylor complained of his "strong propensity to argument."[63] Not everyone lived up to Scott's high expectations, but only Nelson let him down in this manner. In effort if not impact, the expedition's other biological work paled in comparison with Wilson's penguin studies—and Scott prized effort above all.

Ultimately, the Crozier party had returned with three eggs for their extreme exertion. The eggs were frozen, of course, and their contents proved of little value. The theory that inspired the Winter Journey was soon discarded. While individual embryos might contain evidence of a species' history under the Lamarckian model of evolution that Haeckel and many other nineteenth-century zoologists accepted, they need not do so under the neo-Darwinian synthesis that gained ascendency in the twentieth century. Still, a paradigm shift does not lower the scientific status of prior research. Wilson expressed his deeply religious view of the trip's scientific purpose in one of his contributions to the expedition's newspaper:

> THE Silence was deep with a breath like sleep
> As our sledge runners slid on the snow,
> And the fate-full fall of our fur-clad feet
> Struck mute like a silent blow . . .
> AND this was the thought that the Silence wrought
> As it scorched and froze us through,
> Though secrets hidden are all forbidden
> Till God means man to know,
> We might be the men God meant should know
> The heart of the Barrier snow.[64]

Cherry-Garrard never fully recovered psychologically from the Western Journey and suffered from clinical depression for the rest of his life. Possessing deep inner resources, Wilson and Bowers bounced back physically and emotionally—but this merely served to qualify them for Scott's upcoming Polar Journey.

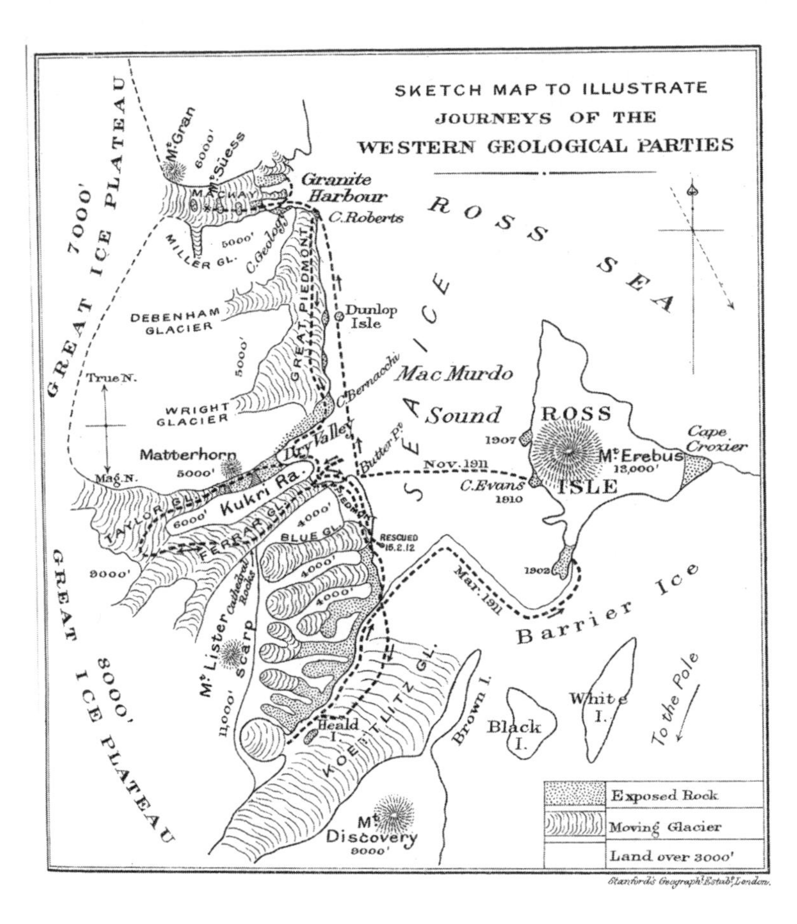

Western geological journeys of the *Terra Nova* expedition (1911–12), from *Scott's Last Expedition* (New York, 1913).

CHAPTER 7

Discovering a Continent's Past

ALTHOUGH THE ENGLISH HAD LONG VALUED CERtain substances from the earth, such as minerals, peat, and coal, rock collecting was more of a hobby than a science in Britain before 1800, and fossils carried little meaning. A century before *Discovery* sailed, most British naturalists assumed that organic species, each perfect in its original creation, never died out. Fossils represented mere sports of nature or still living types. They offered no clues to the past except perhaps as evidence of the global reach and deadly impact of the biblical Deluge. On the basis of the Genesis account, many British naturalists believed that the earth was only about six thousand years old, with its basic features carved either in the beginning by God or later by the Noachian flood. Geology was less a systematic study of nature than a devotional lesson in the supernatural. It offered scant justification for exploring a lifeless, icebound southern continent. Nature was static and Antarctica, if it existed, an eternal wasteland.

A series of nineteenth-century conceptual revolutions in geology and paleontology gave potential meaning to Antarctic research. Early in the century, fieldwork by the French naturalist Georges Cuvier showed that the earth was far older than once thought, its features and conditions had changed over time, and various species came and went in the past. He laid bare the geological column: layer upon layer of sedimentary rock, with each layer containing a characteristic mix of fossilized species. For his findings, he received titles in France and honors from abroad. Across Europe and North America, universities established new positions in the earth sciences. A cadre of midcentury British geologists, including the RGS's Roderick Murchison, extended Cuvier's work by tracing the broad contours of the rock strata and dividing geologic history into named eras and epochs. The empire became their laboratory. Continuing discoveries of prehistoric species fueled popular interest in the field, with the identification of the first upright-walking dinosaurs by British comparative anatomist Richard Owen in 1841 turning paleontology into a Victorian sensation. By the 1850s, queues formed to view his reconstructed giants at the British Museum.

As interest grew, revolutions in the earth sciences began building on themselves. The theories of the British barrister-turned-geologist Charles Lyell elevated current seismic events, such as volcanic eruptions and earthquakes, from merely local, transitory acts of God or nature to mighty engines of a global geologic cycle that continuously renewed the earth's surface through uplift and erosion. The Swiss-American naturalist Louis Agassiz dramatically demonstrated the extent of past climate change by finding evidence of former glacial action even in temperate and tropical locales; and he suggested that ancient ice ages had caused the mass migrations and extinctions reflected in the fossil record. Supported by governments that suddenly saw economic value in such knowledge, scientists in Europe and America began mapping the earth's substratum. And as the theory of evolution gained acceptance late in the century, comparative paleontology became a means to study biological relationships and the former distribution of land, water, and ice. Antarctica no longer looked so distant, different, or irrelevant. It could once have been like Europe; Europe could once have been like it.

* * *

Geology was behind only terrestrial magnetism and geographic discovery as a scientific justification for the *Discovery* expedition. Promoting the venture during the 1890s, for example, *Challenger* naturalist John Murray—whose own main interest was oceanography—played up its potential for geology. "Is there a sixth continent within the Antarctic circle," he asked in the 1894 address that launched the campaign for a British Antarctic expedition, "or is the land nucleus, on which the massive ice-cap rests, merely a group of lofty volcanic hills? This is a question still asked and answered differently by naturalists." The evidence for an Antarctic archipelago came from Robert McCormick, the naturalist on James Ross's 1839–43 expedition who mistakenly characterized all of Victoria Land's mountains as volcanic, and from other explorers who reported finding only volcanic rocks on islands south of Cape Horn. All of this supported the idea of an oceanic archipelago of volcanic islands. During the 1880s, however, the *Challenger* expedition dredged up bits of sandstone, limestone, and other types of continental rock from the Southern Ocean's floor. "These lithological types are distinctly indicative of continental land," Murray noted. The rocks could only have come from melting Antarctic icebergs. Other expeditions made similar finds, which pointed toward the existence of a great southern landmass. An Antarctic expedition, Murray stated, could resolve this question.[1]

In his address, Murray raised a range of issues that would occupy Antarctic geologists for generations. He called on them to survey the rock strata, map the terrain, gauge the depth of the polar ice sheet, look for fossils from more temperate times, explore the interior, and measure the earth's curvature in high latitudes. "All this should be the work of a modern Antarctic expedition," he declared. Speaking after Murray, the Duke of Argyll extolled the proposed expedition's scientific importance, "especially to the science of geology." For his part, Markham wrote, "The main object of the expedition, then, would be to explore the Antarctic continent by land, to ascertain its physical features, and above all to discover the character of its rocks, and to find fossils throwing light on its geological history." Like Murray, Argyll and Markham had no interest in a mere dash to the South Pole.[2]

The expedition's geological goals crystallized in 1900, after organizers named British Natural History Museum geologist J. W. Gregory to head *Discovery*'s scientific staff with full authority on shore. Gregory had led the first geological expedition to Africa's Great Rift Valley and conducted field research in the Arctic. Eminently qualified to direct the *Discovery*'s shore parties, he proposed an ambitious program for the expedition that put geology virtually on a par with terrestrial magnetism as its principal activity. As he outlined it, the explorers would pursue three main topics of research in the earth sciences.

The first topic involved the nature and extent of Antarctic land. Fridtjof Nansen and some other polar experts argued that the Antarctic was much like the Arctic—with a large ice sheet grounded on an archipelago of islands. Following Murray, however, Gregory believed that it consisted of a true continental landmass, though it might be partly submerged. "There is," he explained, "little doubt that Antarctica is geologically a continent, consisting of a western plateau, composed of achaean and sedimentary rocks like those of Australia, and of an eastern volcanic chain. But whether Antarctica is still a continent geographically is less certain; and this question can only be conclusively settled by a survey." Moreover, he proposed investigating whether Victoria Land's coastal range represented a southern extension of the mountains that ran along the Pacific Ocean's western rim, and whether it linked in the south to the mountains of the Antarctic Peninsula, which were widely viewed as an extension of the Andean range. "In that case," he wrote, "the great tectonic lines which bound the Pacific to the east and west are connected across the Antarctic area; and if that can be proved the unity of the great Pacific depression will be completely established."[3]

In addition, Gregory planned to take regular geodetic and seismic readings. No previous Antarctic expedition had tried to make such measurements. "The principal geodetic work of the expedition," he wrote, "will be the continuation of the line of gravity determinations that has now been carried from California across the Pacific to Sydney." Gregory arranged with the Australian instrument maker and former government astronomer Robert Ellery to supply three pendulums for the expedition's use in making gravity readings. With these, Gregory hoped to resolve the earth's precise oblate ellipsoid shape, which was still unknown for the south. "The International Geodetic Commission had expressed their conviction that a gravity survey in the region would be of the greatest benefit to higher geodetic theories," German polar scientist Georg von Neumayer told the Royal Society's 1898 Antarctic meeting.[4] *Discovery* also carried a seismometer and duplex recorder to take the first sustained readings of the earthquakes from an Antarctic reference point.

The hunt for fossils stirred the most interest. Virtually none were then known from south of the Antarctic Circle, but because of recognized similarities in the past flora and fauna of the three inhabited southern continents, many scientists believed that Antarctica was once warm enough to sustain life and serve as a link for the distribution of species. To account for the seemingly abrupt, widespread appearance of higher plants in the fossil record of the known continents, for example, Charles Darwin wrote to the former Ross Expedition botanist Joseph Hooker in 1881, "I have sometimes speculated whether there did not exist somewhere during long ages an extremely isolated continent, perhaps near the South Pole" where those plants gradually evolved. Having no conception of continental drift or plate tectonics, geologists postulated receded oceans and ancient land bridges connecting the various southern continents through the polar region. Gregory argued that fossils in Antarctica "alone can settle the problems of zoological distribution in South America, South Africa and

Australia." Further, fossils were then the only means to date rock formations and reveal a region's geologic history. For these reasons, Britain's leading expert on biodiversity, Zoological Society Secretary P. L. Sclater, hailed paleontology as "the most promising field" of Antarctic exploration.[5] In 1899, British Museum (Natural History) Director Ray Lankester made much the same claim in lobbying for government support of the *Discovery* expedition.

The six-hundred-page *Antarctic Manual* assembled by the British Museum's George Murray for the *Discovery* expedition devoted five of its eighteen chapters to these three topics. With respect to gathering rocks, the *Manual* warned, "Rock-specimens collected here and there, without correlative information as to the general structure of the district, are rarely worth the trouble and expense of transport." Only a systematic regional survey, with specimens from all exposed strata, held much meaning. Sedimentary strata offered the most potential, the *Manual* stressed, because they held clues to the climate and distribution of land and sea in earlier ages. "The most useful information on these questions may be afforded by remains of animals and plants found imbedded in the rocks," the *Manual* explained, particularly the discovery of *Glossopteris* plant fossils of the type common to Africa, Australia, India, and South America in the late Paleozoic era. Finding *Glossopteris* fossils in Antarctic rocks from that era would support the theory that these four regions were once linked through Antarctica. If these fossils were also found in older rocks, the *Manual* noted, it would show that prehistoric global cooling had allowed these presumably cold-climate plants to move north from the Antarctic.[6]

The expedition's ambitious geology program largely collapsed after the power struggle between the RGS and Royal Society led Gregory to resign his post as science director, less than three months before *Discovery* sailed. When naming Gregory to the position, a Joint Committee of the two societies had promised him control over scientific research and suggested that Scott and the *Discovery* vessel would not remain south over the winter but would simply deposit and retrieve Gregory's shore party. Instead, Clements Markham and the RGS insisted on modeling the venture after earlier Royal Navy expeditions to the Arctic, with Scott in command and *Discovery* wintering on-site. Despite the Royal Society's strong representation on the Joint Committee, the RGS had raised most of the private funds for the mission, and Markham used this leverage to get his way.

The wrangling went on for months. Edward Poulton, a leader of the Royal Society faction within the Joint Committee, made an impassioned appeal on Gregory's behalf in the journal *Nature*, arguing that the embattled science director's "chief subject was Geology, a science which pursued in the Antarctic Continent would almost certainly yield results of especial significance."[7] Poulton's pleas fell on deaf ears within the RGS and failed to unite the Royal Society. The RGS cared most about geographical discovery; an influential bloc within the Royal Society cared most about terrestrial magnetism. In the end, the expedition's final instructions gave equal emphasis to these two activities while making only two passing references to geological research and saying nothing about fossils. They placed Scott in complete command, with authority to winter the ship in the Antarctic. Stripped of control, Gregory quit. The expedition's physicist, William Shackleton, who had supported Gregory's position that scientists rather than sailors should control the mission, was promptly sacked.

Responsibility for geodetic and seismic studies fell on Shacketon's replacement, Louis Bernacchi. A veteran of Borchgrevink's *Southern Cross* expedition, Bernacchi had hands-on experience taking magnetic readings in the Antarctic but little knowledge of how to use the geodetic and seismic equipment that accompanied the *Discovery* expedition. Some items broke in transit; others never operated properly. Bernacchi set up the equipment in huts at winter quarters and took regular readings, but researchers who later analyzed the data found them wanting. Worse problems bedeviled the expedition's efforts at field geology.

With his new ship scheduled to sail in scarcely ten weeks and countless other concerns calling for his attention, Scott struggled to find a geologist to replace Gregory. In light of what had happened, no experienced scientist stepped forward on such short notice. Scott eventually settled on twenty-two-year-old Hartley Ferrar, who had graduated with second honors in natural science from Cambridge one month before his appointment to the expedition's staff. It was neither an inspired nor an inspiring choice. By all accounts, Ferrar had spent more time playing sports at college than studying science. Aloof and withdrawn, he was treated harshly by the ship's officers on the *Discovery*'s outbound voyage. Early on, Scott privately dismissed Ferrar as a "conceited young ass." Markham described him as "very young, very unfledged, and rather lazy."[8] In the hazing ceremonies that marked the occasion of *Discovery*'s first crossing the equator, the sailors subjected Ferrar to particularly rough treatment.

Problems persisted after *Discovery* reached winter quarters. Ferrar kept to himself during the first winter, either holed up in his cabin or poking around the vicinity, which he quickly found uninteresting. "The first thing that strikes one here is the ordinary appearance of the rocks," he wrote in June 1902, depicting the Hut Point peninsula as "four square miles of bare rock, entirely of volcanic origin"—mostly basalts and trachyte.[9] His reports identified five volcanic "craters" or cinder cones in the area, the largest of which he called Observation Hill, but dwelt on the sameness of the rocks. "No particular sequence is indicated by them," Ferrar complained. "The bare land-surfaces are usually covered to a depth of six inches by a loose cloak of rock-debris" over a permanently frozen base. "Here decomposition and disintegration proceed simultaneously, and any particles loosened by frost from the upper surfaces are at once blown away by the wind," leaving no clue to the region's past.[10] The rest of Ross Island, he wrote, was more of the same, and he expressed little interest in exploring it. Instead, Ferrar looked across McMurdo Sound to the mountains of Victoria Land, where he hoped to find rock strata with sedimentary layers. There, he wrote during the first winter, "specimens of any rocks in 'situ' whether igneous or sedimentary should be collected. If we are lucky enough to find the latter, a careful search for fossils, and a diagrammatic sketch should be made."[11]

Although Ferrar accompanied two trips to Victoria Land during the next sledging season, the results were disappointing. The first trip left on September 11, 1902, under the command of Albert Armitage. "My orders were: To proceed to the westward and make a reconnaissance, the primary object being to discover, if possible, a practicable route to the interior of South Victoria Land," Armitage wrote. "I was to give Ferrar every possible opportunity of studying the geology of the country."[12] Within a week, however, Ferrar developed scurvy, which hindered his ability to conduct research and forced the party to return early without finding a glacial pass leading inland. On one occasion, Ferrar became so exhausted that he collapsed on the ice and but for the aid of a companion would have frozen to death in −45°F weather. Armitage deemed Ferrar unfit for sledging.

Armitage launched his main assault on the western mountains on November 29, 1902, but this time he relegated Ferrar to a support party that never went beyond the foothills. The goal of this Western Sledge Journey was to cross over the mountains to find what lay beyond and perhaps reach the Magnetic South Pole. Without any certain idea of the route and again impeded by scurvy, the main party struggled up first one way and then another to the verge of the Polar Plateau before turning back without venturing onto it.

While Ferrar's support party turned back early in the journey and found little new, the main party made some promising observations and collections high in the mountains. "Mr. Armitage's pioneer journey though the mountains proved the existence of plateau features as well as horizontal structure," Ferrar noted. "The specimens he brought back included a sandstone which . . . suggested the probability of the existence of fossiliferous sediments in the district." The sandstone layer lay above the mountain's gneiss base and broad granite shoulders at up to eight thousand feet, Armitage reported, and was topped by a dolerite cap. Reaching this sedimentary formation, which Ferrar called Beacon Sandstone for the Beacon Heights region where it was first found, became the young geologist's objective for the next year. "Among the questions still to be solved," he noted in August 1903, "is, what are the ages of these rocks?" His enthusiasm for such questions and acceptance of a supporting role impressed Scott. By the end of the sledging season, Scott described Ferrar as "a changed youth:" "He was objectionable, he is now a non entity and knows it, therefore in time he will be an acquisition to our little band."[13]

Ferrar came into his own as a field geologist during the expedition's final summer. After Armitage's Western Sledge Journey fell short, Scott tried the route himself in November 1903: up the narrow, icy pass that became known as Ferrar Glacier to its nine-thousand-foot summit, and then charting new ground across two hundred miles of the Polar Plateau. Ferrar went along as far as a sandstone outcrop near the summit. He then split off with one sledge and two sledgers, Petty Officer Thomas Kennar and Seaman William Weller, to gather rock specimens and look for fossils. Scott proceeded west with another sledge and five men. Both parties made Antarctic history.

Because of a severe blizzard that kept the combined team in its tents for seven days at over seven thousand feet, Ferrar's party had only one month's rations when it finally separated from the main group on November 11. Scott's party had rations for a longer trip, but also farther to go. "If I were asked to name the most miserable week

I have ever spent," Scott later wrote of this enforced encampment, "I should certainly fix on this one. Throughout the whole time the gale raged unceasingly. . . . In our tent we had one book, Darwin's delightful 'Cruise of the "Beagle,"' and sometimes one or another would read this aloud until our freezing fingers refused to turn the pages." He named the place "Desolation Camp." It stood beneath an immense icefall, on a small patch of hard snow that the men had found on the bare blue ice at the onset of the blizzard.[14]

Both parties needed to get back to the ship by the end of December to help free it from the sea ice for the return journey to Britain. With little time and much to do, once the blizzard finally ended, the parties set off rapidly in opposite directions. Ferrar's party man-hauled a jury-rigged, seven-foot sledge that showed the wear of three sledging seasons, broke down frequently, and pulled heavily over rough surfaces due to damaged runners. Scott's eleven-foot sledge handled only slightly better. The weather was bitterly cold and often windy. "Luck was not with us this trip," Scott wrote in his diary shortly before the parties split, "and yet we have worked hard to make things go right."[15]

From Desolation Camp, Ferrar took a circuitous route down the glacier looking for fossil-bearing sandstone. "I was quite astonished to learn the number of places he had visited and the distances he had traversed in pursuit of his object," Scott later wrote. "For each specimen of rock which Ferrar brought back was obtained only by traversing long miles of rough ice, by clambering over dangerous crevassed slopes, and by scaling precipitous cliffs."[16]

The best finds came from near the top of the sandstone column. In a glacial moraine visited on the first day, Ferrar unearthed detached sandstone blocks darkened by carbonaceous matter suggesting organic origin. "These were our first evidence of Antarctic life in the geological past," he noted, but their charred condition revealed little about the character of that life and nothing about the rocks' age. The next day, the party reached an outcropping of sandstone capped with overlaying dolerite. "Imagine my delight," Ferrar wrote, "when, arriving with bag and hammer at the rock face, I found thin, black, irregular bands in the pure white sandstone." These bands again hinted at past life without showing distinct fossils or telling the formation's age.[17]

Hoping for better finds deeper in the column, the party worked its way down the glacier to lower beds of the sandstone strata, which Ferrar found notably undisturbed but frustratingly devoid of fossils. "Only one of its horizons contained organic remains," he reported, "and these of a most doubtful nature": cylindrical casts in the stone from six to thirty-six inches long and half an inch wide. Indistinct, wormlike markings that might or might have organic origin appeared here and elsewhere in the stone.[18]

Increasingly plagued by snow blindness during the trek, Ferrar nevertheless completed the first credible geological survey of the central Victoria Land mountains. Despite ice and snow covering most of the terrain, by observing exposed mountains, cliffs, and outcrops at various altitudes he managed to chart the relative levels and locations of various rock types. In the main, Ferrar found gneiss, metamorphic limestone, and schist at the base, overlaid by granite, then sandstone with a dolomite top. He did not know it, but he also secured the first recovered fossil showing Antarctica's ancient link to other southern continents. The official report on Ferrar's specimens, published in 1907, concluded that they held no certain evidence of organic agency. Twenty-one years later, a British Museum paleontologist in London cracked one of them open to find a fossilized impression from a *Glossopteris* plant—just what Ferrar had been seeking. In his last season's work, Ferrar had salvaged something for geology, but as Scott biographer David Crane wrote, "there was a strict limit to what a young and inexperienced geologist plagued by snowblindness and operating with a broken-down sledge and two seamen to assist him could do in less than a month's exploration."[19]

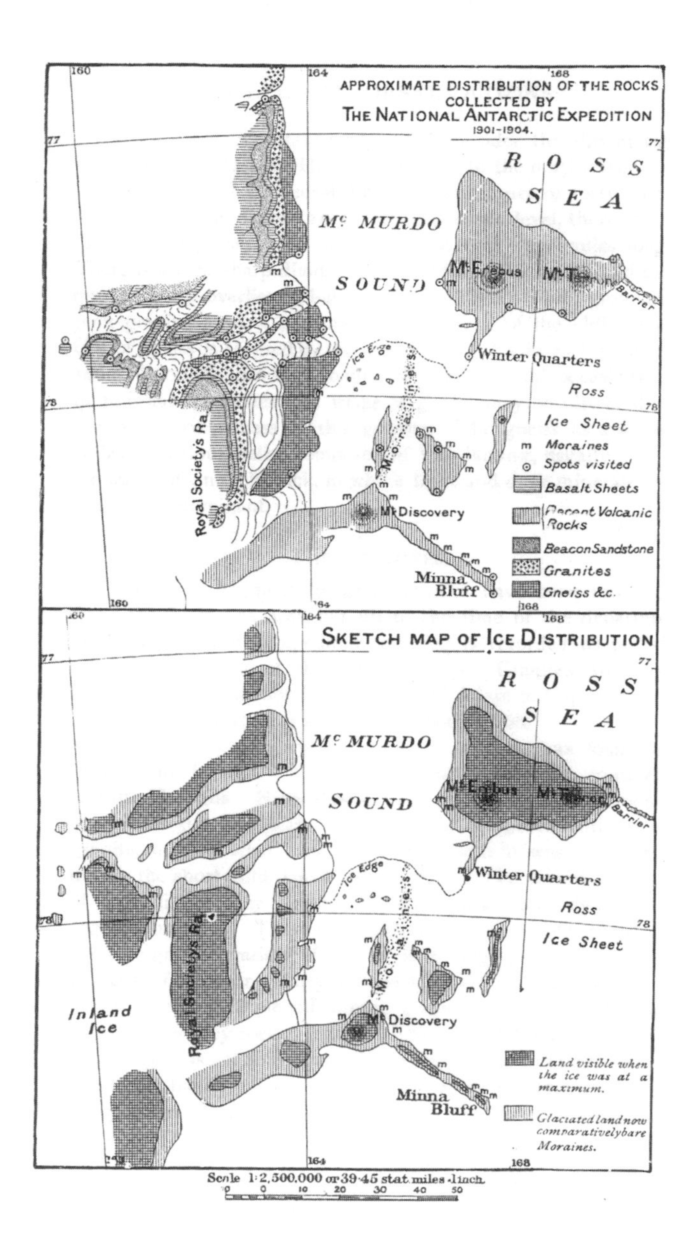

Map of the exposed geological formations around McMurdo Sound by geologists on the *Discovery* expedition, from Robert Scott's *Voyage of the "Discovery"* (New York, 1905).

* * *

As Ferrar's party worked its way down the glacier from Desolation Camp, Scott's group dashed as far west over the Polar Plateau as possible in the six weeks left for its trek. After ten days, Scott reduced the party to himself and the two strongest sledgers—Petty Officer Edgar Evans and Royal Navy Stoker William Lashly—and sent the others back. "With these two men behind me," Scott wrote, "our sledge seemed to become a living thing, and the days of slow progress were numbered."[20] They sped across the frigid, windswept, trackless terrain, stopping only for meals and sleep. Traveling rapidly for a month over a continental ice sheet, where no rocks, mountains, or crevasses disrupted the icy scene, offered no opportunity or occasion for studying geology. On the way back, however, after regaining the coastal mountains and starting their descent, Scott's party literally dropped into the most significant geological find of the heroic age of Antarctic exploration.

In a glacial basin over halfway down the mountains at an altitude of about 2,500 feet, Scott's party came to what looked like a fork in the Ferrar Glacier. The party had ascended the glacier's eastern arm, but the northern arm looked wider. It too must descend to the sea, Scott reasoned, and perhaps was the glacier's main outlet. "At this point I had determined to do a small piece of exploration," he wrote. With the sledge and all the gear, on December 17, the men started down the northern arm, which quickly became steeper than expected. By evening, what Scott described as "a lofty groin of rock" appeared ahead and seemed to block the glacier's path. The next morning, the men proceeded without the sledge to see what lay before them. The ice became so steep and disturbed that they roped up for safety. Streams of meltwater flowed over the surface of the dwindling glacier. "We descended into one of these watercourses and followed it for some distance, until, to our surprise, it came to an end, and with it the glacier itself," Scott observed in his diary. "Before us was a shallow, frozen lake into which the thaw-water of the glacier was pouring." They climbed down the glacier's snout into a long, narrow, snow-free valley filled with glacial till and some muddy silt. "What a splendid place for growing spuds!" Lashly exclaimed.[21]

No one had ever seen anything like this in Antarctica: an interior dry valley. Here, anyone could see the land that lay beneath the continent's ice and snow. Over the next century, it would become the principal place to study continental geology in the Antarctic. At the time, however, Scott simply marveled at the panorama. Towering cliffs adorned by hanging glaciers surrounded a silent, seemingly lifeless landscape of rocky hills, sandy beaches, flowing streams, and a series of lakes, some of them ice-free. "I was so fascinated by all these strange new sights that I strode forward without thought of hunger until Evans asked if it was any use carrying our lunch further," Scott wrote. "We all decided it wasn't, and so sat down on a small hillock of sand with a merry little stream gurgling over the pebbles at our feet."[22] After lunch, the men walked over five miles down the valley to where it was nearly blocked by a groin of boulders. They climbed the rocks in hopes of sighting the sea but saw only the valley extending on toward another groin. So they returned the way they came and eventually reached the ship on Christmas Eve.

Scott never forgot what he saw that day. "I cannot but think that this valley is a very wonderful place," he wrote in his diary. "We have seen to-day all the indications of colossal ice action and considerable water action, and yet neither of these agents is now at work."[23] He knew the valley deserved close study and would send geologists back for that purpose, but it would have to wait for another expedition. With the midnight sun finally melting the sea ice, the men focused all their efforts on freeing the ship and heading for home.

Although Ernest Shackleton left Antarctica before Scott's discovery of the Dry Valley, he soon heard about it and was struck by the attention it received. In announcing his *Nimrod* expedition, Shackleton pledged "to continue the biological, meteorological, geological, and magnetic work of the *Discovery*."[24] When his expedition ultimately wintered directly across McMurdo Sound from Ferrar Glacier, geologizing in the Dry Valley joined his list of objectives along with reaching the geographic and magnetic poles, climbing Mount Erebus, finding fossils, and surveying the region generally. He knew that geological discovery represented an important part of a successful Edwardian expedition.

Organizing his effort rapidly and on a shoestring, however, Shackleton initially picked a geologist even less qualified than Ferrar. Only twenty, Raymond Priestley had completed just two years of college at Bristol when he interviewed for the post. "I was not academically qualified," he later admitted, but he did answer one question right: when asked, "Would you know gold if you saw it?" he said yes. That was good enough for Shackleton.[25] But Shackleton's touting of the expedition attracted the attention of the University of Sydney's eminent and energetic geology professor, Edgeworth David, and David's former student, University of Adelaide mineralogist Douglas Mawson. Having conducted extensive field studies in Australia, both men knew the region's geology and how to work in harsh terrain. With David's help, as Shackleton passed through Australia on his way south, he gained funding from the Commonwealth government, and he added both David and Mawson to the expedition's scientific staff. Together with Priestley, whom they mentored, David and Mawson gave Shackleton the finest team of geologists yet assembled in the Antarctic. All three went on to illustrious careers in science and public service, and were later knighted. They saw much in Antarctic rocks and landforms that Ferrar missed.

Their discoveries began at Ross Island, which Ferrar had found uninteresting. On Mount Erebus and elsewhere, David indentified large quantities of a rare type of volcanic rock that Ferrar scarcely mentioned—kenyte, which *Discovery*'s original science director, J. W. Gregory, had first found in East Africa. The predominance of kenyte on Erebus, David observed, suggested that it was a continental rather than oceanic volcano and thus buttressed the emerging view of Antarctica as a continent. Ferrar had failed to recognize the extent of kenyte on Ross Island because, as he reported it, "Owing to the difficulties of access, few rock-specimens could be obtained from Mount Erebus itself."[26] Within two months of reaching Ross Island, in contrast, David, Mawson, and four others had climbed to the mountain's summit and surveyed its crater. Although the party made this ascent mainly for meteorological purposes, David and Mawson collected rock samples and made geological observations along the way.

Further, where Ferrar had reported finding "no particular sequence" in Ross Island's volcanic rocks, David saw otherwise. Based on a close examination of intrusions and layering, he wrote, "We are now in a position to say that, on the whole, the trachytes appear to have been the oldest rocks, the kenytes to be of intermediate age, and the basalts the newest."[27] Among the basalts, olivine basalt came first, then black basalt. He also indentified some modern kenyte. He made the island's geologic history come alive for Shackleton and others on the expedition, many of whom became avid rockhounds, gathering specimens throughout the region.

The richest reward from David's proselytizing came when, on its race to the South Pole, Shackleton's Southern Sledge Party, which included no geologists, discovered a continuation of Victoria Land's Beacon Sandstone formation near the summit of Beardmore Glacier at latitude 85° south. It also found a seemingly older limestone series there and brought back rock specimens containing fossils of tiny warm-water invertebrates, *Archaeocyatha,* from the early Cambrian period. Even more revealing of the region's geologic history, near the top of the sandstone formation, the party found seven seams of coal. The men stopped to collect samples, including a sandstone fragment with a thin black band that, on later study, appeared to come from the stem of a coniferous tree. "This is the first recognizable fossil plant that has been obtained from the Victoria Land area of the Antarctic," David wrote. These fossils, he concluded, placed the limestone in the Cambrian period and the sandstone probably in the Upper Devonian or Lower Carboniferous period. The coal appeared to date from the Permian. "The discovery of coal and fossil wood has a very important bearing on the question of the past geological history of the Antarctic continent," Shackleton boasted at a meeting of RGS members on his return. In finding them, he had bested their man Scott.[28]

The Southern Sledge Party, which carried the hopes of the entire expedition, consisted of only four persons. Ten others remained behind. In consultation with David, who served as the expedition's science director, Shackleton ordered a rich program of research and exploration for them. In accord with his instructions, David led a three-man Northern Sledge Party to the South Magnetic Pole. After crossing McMurdo Sound to Butter Point near the foot of Ferrar Glacier in early October, this party followed the west coast of the Ross Sea north for 200 miles to Larsen Glacier, then ascended Victoria Land's mountains and traversed the Polar Plateau for 250 miles to the magnetic pole. As the expedition's physicist in charge of magnetic readings, Mawson naturally joined this party, which, because of his and David's presence, took extensive geological notes and collected rock samples while conducting its magnetic survey. They found that the Beacon Sandstone formation, together with underlying limestone, extended north at least as far as latitude 75° south.

The Northern Sledge Party's main geological work, however, was supposed to occur at the end of its journey. After conducting the magnetic survey and attaining the magnetic pole, it was to return south to Butter Point by the first week of January and join a three-man Western Sledge Party consisting of Priestley, Philip Brocklehurst, and expedition pony-trainer Bertram Armytage but no ponies. The draft animals had all gone south with Shackleton. Hauling their own sledges, the two parties would jointly explore the nearby mountains and Dry Valley under David's supervision. Shackleton gave David explicit directions on these points. "I particularly wish you to be able to work at the geology of the Western Mountains, and for Mawson to spend at least one fortnight at Dry Valley to prospect for minerals of economic value on your return from the north," he wrote. "I consider that the *thorough* investigation of Dry Valley is of supreme importance."[29] He gave corresponding instructions to the Western Party, which was to leave Cape Royds in early December, conduct initial fieldwork in the Ferrar Glacier region, and then rendezvous with the Northern Party at Butter Point during the first week of January for the first systematic study of the Dry Valley.

The schedule allowed time for Priestley, Brocklehurst, and expedition biologist James Murray to explore the northern flank of Mount Erebus in late November, before the Western Sledge Party departed. No scientist had examined this region, which included the mountain's oldest main crater, and David thought it might prove interesting. Accordingly, Shackleton left instructions for "a geological reconnaisance" of the area. Five men set out on this effort with a week's supply of food, Shackleton later reported, "but carried only one tent, intended to hold three men, their idea being that one or two men could sleep in the bags outside the tent."[30] This arrangement left Priestley sleeping outside and four others in a three-person tent when a ferocious blizzard hit on the first night.

The blizzard pinned the party in place on a glacier that sloped down toward rocks and the sea. "Inside the tent for the next three days we were warm enough in our sleeping-bags . . . [t]hough we could not cook anything," Murray wrote. They survived on dry biscuits and raw pemmican. "The little snow under the floorcloth was squeezed in the hand till it became ice, and we sucked this for drink." Outside, wrapped only in a sleeping bag and discarded tent cover, laying abreast of a howling wind that threatened to roll him down the slope, Priestley was worse off than the others. "I then tried the experiment of lying head to the wind," he wrote. "It was in this position that I spent the next seventy-two hours, getting shifted down a yard or two at a time at every change in the direction of the wind." The others passed him food, but he kept sliding farther from the tent, and no one could see in the storm or stand against the wind. "It may sound like an exaggeration," Priestley wrote of their position, "but it must be remembered that we were lying on the slopes of a clean-swept glacier. . . . A slip on the ice meant very serious danger of destruction." On the bare glacier, Priestley did not even have snow. "For nearly eighty hours I had nothing to drink but some fragments of ice that I could prise up with the point of a small safety-pin." When the storm finally subsided, the party raced back to the hut without conducting any further research.[31]

Having survived their ill-fated reconnaissance, Priestley and Brocklehurst set out on December 9 with Armytage for the Ferrar Glacier to begin their Western Sledge Journey. "The main object of our journey up the Ferrar Glacier was to examine the Beacon sandstone at any accessible exposures with a view to the discovery of any traces of former organic life," Priestley explained.[32] They initially aimed for exposures near the glacier's summit, but the midnight sun had made the surface too soft to traverse in the time available. Instead, the party stopped midway up the glacier, where Ferrar's chart showed another exposure. "The whole of the bluff opposite is marked as Beacon sandstone, and from the face of the cliff here it is easily seen, for at least 3000 feet, to be granite," Priestley railed. The only accessible sandstone was at his feet, in lateral moraines that carried debris from higher elevations. "The sandstone is very weathered, dropping to pieces in many cases with a single blow," he complained. "I am faced with the necessity of examining for fossils rocks which I should carefully avoid if I were at home or anywhere else. I have never seen a sedimentary rock that looked more unfossiliferous."[33] Giving up on fossils, Priestley took notes on geology and turned back in time to reach Butter Point by January 1 for the planned rendezvous with David.

Of course, the Northern Party never arrived. On January 1, it was still more than 150 miles from the South Magnetic Pole. Man-hauling heavy sledges on half-rations in extreme conditions, David and his men would not reach the pole and regain the coast until February 2, by which time the sea ice had gone out and they could not proceed south toward Butter Point. Stranded on the coast, they prayed for *Nimrod* to rescue them. Meanwhile, following Shackleton's orders, the Western Party waited for David on the sea ice near Butter Point for nearly a month. During this time, the party made only two brief excursions: one to the eskers or "stranded moraines" at the snout of the nearby Blue Glacier and one to the lower end of the Dry Valley, which opened to the sea a few miles northeast of Butter Point.

On January 24, one day before the Western Party was authorized by Shackleton's instructions to leave Butter Point, the sea ice under its tent broke loose and drifted two miles from shore. "The position seemed to be rather serious," Armytage later reported to Shackleton, "for we could not hope to cross any stretch of open water, there was no reasonable expectation of assistance from the ship, and most of our food was at Butter Point." Killer whales began butting against the ice floe. "It is a well-known fact that the killer-whale lives round about the pack," Priestley noted, "and breaks it up by bumping it in order to get the animals off it for food and outside our tent there were a large school of them playing and one of them bumped directly beneath our tent cracking the ice in all directions." After eighteen hours of terror, the floe collided with a piece of fast ice. "Not more than six feet of the edge touched, but we were just at that spot, and we rushed over the bridge thus formed," Armytage reported. "We had only just got over when the floe moved away again, and this time it went north to the open sea."[34]

Nimrod rescued the Western Party at Butter Point a day later and on February 4 found David and his men on the coast near the foot of Larsen Glacier. It was a close call for both groups, but Shackleton managed to return home without losing anyone—something Scott could never claim. After picking up Shackleton's Southern Party at Hut Point and its geological specimens at nearby Cape Armitage on March 4, *Nimrod* steamed hastily north ahead of the fast-closing winter. It left behind rock samples from the Northern Sledge Journey's outbound leg that David had stored on Depot Island.

Scott returned to the Antarctic on *Terra Nova* in 1910, committed to reaching the South Pole but determined not to neglect the Dry Valley. After his Polar Journey and Wilson's Winter Journey to Cape Crozier, geology and the Dry Valley figured most prominently in his plans. "I have arranged for a scientific staff larger than that which has been carried by any previous [Antarctic] expedition," Scott explained early in 1910, and "have regarded geology as one of the most important interests which can be served on our expedition." Accordingly, he named three geologists to the staff. To lead them he recruited Griffith Taylor, one of David's former students from Australia then studying at Cambridge. Passing through Sydney on his way south, Scott added two more of David's students, Frank Debenham and *Nimrod*'s Priestley, who had followed David to Australia after returning from the Antarctic. In deferring to David, Scott hailed him as "probably the best judge in the world of the work which remains to be done and of the men who should be selected to do it."[35] All three geologists were under thirty when *Terra Nova* sailed and were destined for long and distinguished careers in science.

Having three geologists served Scott's original plan for the expedition. He envisioned it having two winter bases: a small one on King Edward VII Land at the Ice Barrier's east end and a large one on

Ross Island near the barrier's west end. The Ross Island base, which Scott placed at Cape Evans midway between *Discovery*'s Hut Point and *Nimrod*'s Cape Royds, would serve as the starting point both for his own journey south to the pole and for excursions west into Victoria Land. He planned to send one geologist in each of three directions—east, south, and west—with the senior one, Taylor, going south with the Polar Party as far as Beardmore Glacier. Apparently because of tensions between himself and Taylor, Scott decided to take the expedition's physicist, Charles Wright, instead of a geologist on his Polar Journey. This left two geologists, Taylor and Debenham, to go west.

The third geologist, Priestley, joined the Eastern Party that was supposed to explore King Edward VII Land, with an eye toward surveying the Ice Barrier's unknown east side. To lead this party, Scott picked Royal Navy Lieutenant Victor Campbell, whom Debenham described as "the best all round" member of the expedition save Wilson.[36] Sailing aboard *Terra Nova* in search of a wintering site, this party came across Amundsen's base at Bay of Whales. Conceding the region to the Norwegians, the Eastern Party became the Northern Party, with its winter base at Victoria Land's Cape Adare, where Borchgrevink's *Southern Cross* expedition wintered in 1899. The party's instructions had listed Smith's Inlet, on the mainland west of Cape Adare, as a back-up wintering site, but it proved inaccessible. Priestley could do little original geological fieldwork from Cape Adare—a narrow spit of land cut off by cliffs and water from the mainland. The cape's geology was well known from collections brought back by Borchgrevink. Campbell objected to wintering there, but after steaming to King Edward VII Land and back, *Terra Nova* was low on coal for the voyage to New Zealand, and its captain insisted on offloading the party. Scott's hopes that it would explore the Southern Ocean coast westward beyond Cape North never materialized.

All in all, Scott had sketched out a highly ambitious plan for science. Certainly the *Discovery* and *Nimrod* expeditions had not attempted anything so bold, and Amundsen did not even offer a pretense of science to mask his polar ambitions. "Doubtless there are those who will criticize this provision in view of its published objective—that of reaching the South Pole," Scott said of the role for science in his expedition. "But I believe that the more intelligent section of the community will heartily approve of the endeavour to achieve the greatest possible scientific harvest which the circumstances permit." In addition to the planned excursions from base, Scott took along personnel and equipment to resume the ongoing scientific observations that had marked the *Discovery* expedition, with Wright taking geodetic and magnetic readings; meteorologist George Simpson keeping weather records; marine biologist Dennis Lillie tow-netting, trawling, and dredging at sea; and invertebrate zoologist Edward Nelson monitoring the marine life and tides in the vicinity of winter quarters. When *Terra Nova*'s third officer, Scott's brother-in-law Wilfred Bruce, met Amundsen and his men at the Bay of Whales, he recognized the difference between them and Scott's team. "They have 120 dogs, and are going for the Pole! No science, no nothing, just the Pole!" he warned. "If their dogs are a success, they are more than likely to be there first."[37]

Tapped to lead the Western Party for two sledging seasons, Taylor opened his account of its journeys by describing its area of operations as seen from the expedition's main base. "As you stand on Cape Evans with your back to the steam cloud of Erebus," he wrote, "you see across McMurdo Sound a glorious range of mountains running due north and south and rising to 13,000 feet in the south-west." He called them the Western Mountains, but they were part of a range then known to run from Cape Adare in the north at least to Beardmore Glacier in the south and later found to extend across the continent to the Weddell Sea. "Their southern limit," he wrote of the mountains in plain sight of Cape Evans, "is the extinct volcanic cone of Discovery, and far to the north one can follow the same range of snow-clad peaks until it merges with the gray line of the horizon." Koettlitz Glacier stood at the base of Mount Discovery in the south; Granite Harbor lay just over the horizon in the north. Scott gave Taylor and Debenham the task of conducting a geological survey of the mountainous region between these two landmarks, working the southern part in the short first summer and the northern part in the long second summer.[38]

Getting to work quickly, Taylor's group headed first for the Dry Valley in the heart of the Western Mountains. *Terra Nova* had reached Cape Evans on January 4, 1911—two weeks after the summer solstice. Just twenty-three days later, the Western Party launched its initial excursion, which lasted seven weeks. Ferried across McMurdo Sound to Butter Point by *Terra Nova* on January 27, the party promptly struck west into the interior. "We were to ascend Ferrar Glacier[,] turn around into Dry Valley and explore it thoroughly," Debenham wrote. "Capt. Scott had entered it from the West [in 1903] and Priestley from the East [in 1909] but no real exploration had been done."[39]

"The object of your journey will be the geological exploration of the region between the Dry Valley and the Koettlitz Glacier," Scott had instructed Taylor at its outset. After surveying the Dry Valley, therefore, the party would head south to the broad Koettlitz Glacier and then return to Ross Island by way of the ice-covered southern end of McMurdo Sound. Because of his interest in the physics of ice, Wright joined Taylor and Debenham on this first Western Journey, which did not overlap with Scott's Polar Journey. Like Wright, Taylor had more interest in glaciers and physical geography than in rocks and fossils. "He left the more orthodox branches of geology to me entirely, including the collection of samples," Debenham noted.[40] Edgar

Evans, who had codiscovered the Dry Valley with Scott and Lashly in 1903, completed the four-person team. They started with a half-ton of supplies and equipment, which they man-hauled on two heavy sledges, but depoted many items, including food, along the way and supplemented their diet with fresh seal meat. With winter coming on, they expected the weather to deteriorate as they traveled and knew that they would add rock specimens to their load.

The party reached the Dry Valley by the customary route—up the Ferrar Glacier to its fork, then down the northern arm. Carefully examining this fork, Taylor determined that two glaciers came together above this point, temporarily uniting "after the fashion of the Siamese twins," and then split apart, with the Ferrar continuing east to the sea and the other turning north to terminate at the Dry Valley. When Taylor later described this dynamic to Scott and explained that the two glaciers did not mix, Scott named the new glacier, and thus its valley, for Taylor. In his diary, Debenham depicted the scenery as "absolutely ripping."[41]

"To our surprise—after five days' pulling over heavy snow in the Ferrar Glacier—we found no snow in the adjoining valley!" Taylor wrote of his introduction to the place that would bear his name. "Imagine a valley four miles wide, 3000 feet deep, and 25 miles long without a patch of snow—and this in the Antarctic in latitude 77 1/2°S." Above the glacier's snout, the valley was carpeted with steep, bare ice sliced lengthwise by meltwater surface streams. "The sledge was almost running away from us," Debenham wrote of the party's steep descent.[42]

After camping for two days above the snout, mapping the valley's upper reaches and collecting rock samples, the party dropped into valley's glacier-free lower part. "As we could not take the sledge beyond the glacier," Taylor noted, "we packed up the tent and sleeping-bags with five days' food and our instruments, and carried them down toward the sea." Lacking backpacks, they lugged their supplies on a strap or pole slung over one shoulder or in a swagman's bundle draped over both front and back. "We looked like a gang of gypsies," Wright observed. He likened Taylor to "a shop Santa Claus hung about with telescope, android, camera, fur gloves and with some grub." Debenham added, "This pack humping down Dry Valley was a new departure in Antarctic work, it generally being considered impossible to go away from the sledges for more than a few hours." The conditions proved ideal for the experiment. "It was warm weather," Taylor noted, "most of the time we spent in Dry Valley—rising sometimes above freezing-point—and everywhere streams were tinkling among the black boulders." Everyone seemed delighted with the place: Wright and Evans frequently broke into song.[43]

The party lingered in the valley for four days, surveying its basic features and fixing its location in relation to known landmarks. The glacier had gouged out the Dry Valley during a period of deep glaciation, Taylor reasoned. The presence of extruded lava on glaciated walls showed that volcanoes had erupted here since the glacier retreated. Granite predominated at the valley's base, with sedimentary layers and dolerite sheets appearing higher up to create a palette of earth tones. The men repeatedly scaled the ridges to gain a wide view or examine side glaciers. Debenham made an extensive collection of geological samples, while Taylor tried his hand at panning for gold. "There were numerous quartz 'leads' in the slates and metamorphic gneisses," he wrote, "which is always promising, and furnishes the 'country rock' of most gold fields. But the quartz was too glistening and pure."[44] While he found no minerals of commercial value, the party was amused by the prospect of a gold rush to Antarctica. Fossils also proved elusive, a fact Taylor blamed on the folded and altered state of the sedimentary rock. On February 7, after traveling nearly to the sea, the men retraced their steps to the glacier, retrieved their sledge, and headed down Ferrar Glacier to Butter Point.

Traveling south and east from Butter Point, Taylor's party explored ice features along the coast, in Koettlitz Glacier's deep delta, and around the Ice Barrier's ragged edge linking Victoria Land to Ross Island. After enjoying idyllic conditions in the Dry Valley, Debenham noted, "We are beginning to experience real Antarctic troubles and we don't find them pleasant." The return trip took longer than expected because the rotten late-summer sea ice on McMurdo Sound forced the party south into nearly impassible regions of sharp pinnacle ice and soft snow mixed with glacial till. "With the big sledge much of the work was a stand and pull game," Wright noted. "The trouble was largely the sand blown over everything—even the heavy snow drifts into which one sinks two feet or so."[45] Evans described it as the worst sledging surface he had ever encountered—and he had been on Scott's brutal 1903 Western Sledge Journey. "It took us seven days to do the twenty miles to Hut Pt," Debenham wrote.[46] When the men tried cutting across patches of intact sea ice, killer whales harried their efforts by battering the floe. With the approach of winter, temperatures dipped far below zero Fahrenheit.

By noon on March 10, Ross Island seemed so close across the sea ice that Taylor spoke of reaching it by nightfall. Open-water leads sent the men south again, however, and into the teeth of a blizzard. When they finally arrived at Hut Point four days later, they found Scott and twelve others waiting for new sea ice to form so that they could cover the remaining twelve miles to Cape Evans after a near-catastrophic trip to depot supplies on the Ice Barrier for the next summer's Polar Journey. Scott's group had lost most of its ponies and nearly some human lives on the rotten sea ice below Hut Point: ice similar to what the Western Party had almost tried to cross. The combined group—seventeen men in all—remained at Hut Point for over a month until stable sea ice reformed. Even then, a blizzard turned what should have been a half-day hike to Cape Evans into a three-day ordeal. When

Taylor blamed Scott for this final foul-up, Scott exploded. Describing the incident as a "very bad fracas," Debenham wrote about Scott, "What he decides is often enough the right thing I expect, but he loses all control of his tongue and makes us all feel wild."[47] Once back at Cape Evans, the Western Party remained there until spring, when Taylor again led men along the Victoria Land coast before leaving the expedition early on *Terra Nova*'s 1912 relief voyage.

Although more ambitious than its first journey because of the area traversed, the Western Party's second journey dealt more with glaciers and physical geography than with rocks and fossils. This time, Petty Officer Robert Forde and Norwegian ski expert Tryggve Gran accompanied Taylor and Debenham. Hauling two heavy sledges, they left Cape Evans on November 14, 1911, crossed the sea ice to Butter Point, and then tracked the coast north on the sea ice for fifty miles to Granite Harbor. "Looking to the north we could see nothing but a great barrier wall of ice along the coast," Taylor wrote about the broad face of Piedmont Glacier, which spilled over most of the shoreline from Butter Point to Granite Harbor. "Except on the capes, no rock was exposed." As a result, he added, "I don't take very full geological notes for obvious reasons—we only see a piece of rock about every three days!" They focused instead on surveying. "We had two separate instruments," Taylor wrote, "a theodolite and a plane-table. With the former I was able to fix far-distant peaks with considerable accuracy, and also by observations on the sun to determine the latitude and longitude of the main stations of our survey. With the plane-table Debenham carried out a unique detailed survey of the coast-line." He hailed this topographical map as the journey's chief achievement in physical geography.[48]

The men reached Granite Harbor on November 27 and settled in for a seven-week stay to study that fjordlike bay in depth. For cooking, sleeping, and storing gear, they built a stone hut on a narrow point of land poking into the bay. "We named the latter Cape Geology, in memory of the chief object of our journey, though we had been able to do very little scientific work so far," Taylor wrote.[49] They surveyed the coast, examined the glacial tongue bisecting the bay, and explored the surrounding granite cliffs and hanging glaciers. Their best chance to geologize came in late December, when they made a ten-day excursion up Mackay Glacier, the massive ice flow that had carved out Granite Harbor. On Mount Suess, a 4,500-foot stranded peak poking through the glacial ice about ten miles inland, they found the rocks they wanted.

"There was an extraordinary mixture of dolerite and sandstone all over," Taylor wrote about the exposed ridge behind Mount Suess, where the party camped. "The sandstone . . . did not look hopeful for fossils. However, there was some shale near the tent, which looked more hopeful. We did not find much beyond worm-casts and ripple-marks at first." He then went off to survey the ridge while Debenham remained behind to examine the shale. By the time Taylor returned, Debenham had hit pay dirt. "He had found some vesicular horny plates," Taylor noted, which later "were identified as the armour-plate of primitive fish, and probably of Devonian age." More searching turned up other fish plates, including two-inch-long pieces with distinct keels that Taylor likened to "the red tiles capping a roof-ridge." These finds added a new dimension to Antarctica's geologic time scale, he explained, "for Cambrian limestones were known, and Permian coal-measures were indicated by Shackleton's specimens. These fish plates identified another set of sediments midway between them." Along the ridge, Debenham also found brown coal.[50]

Following its advance instructions, the Western Party returned to Granite Harbor in early January. Scott had directed *Terra Nova* to pick up the party by January 15 and reposition it farther north on the coast for an added month of fieldwork. The day came and went without any sign of *Terra Nova*. When the ship finally appeared on January 20, it could not push through the late-summer sea ice to pick up the men, and they could not cross the rotten ice to reach it. "We must be an uncommon group of men not to tear out each other's hair in sheer frustration," Gran wrote in his diary for January 28. By month's end, the ship had disappeared and the men began to doubt that it would return. "We had plenty of seal meat and biscuit, but all the other stores were approaching their last week," Taylor noted. He decided to lead his men south, over the Piedmont Glacier, toward Butter Point, where the party had stored supplies and the ship could perhaps reach them. "We took nothing but what we stood up in, and our notes and the instruments," Taylor wrote. If all else failed, the party could carry on along the Ice Barrier's sea edge to Ross Island, as it did the previous year. The grueling trek over deeply creviced glaciers finally ended on February 15, when *Terra Nova* rescued the party some fifteen miles south of Butter Point.[51]

After rescuing the Western Party, the ship sailed north to retrieve Campbell's Northern Party, which it had repositioned to Terra Nova Bay in early January for six weeks of fieldwork following the wasted winter at Cape Adare. At the outlet of several steep glaciers, with moraines rich in debris from the high mountains, the region around Terra Nova Bay was ideal for summer fieldwork but dreadful for a winter camp. Campbell's hapless men experienced both extremes. Within the first month of their arrival, Priestley and Campbell discovered fossil-bearing sandstone rocks in the moraines, including several containing fossilized wood. When the scheduled departure date of February 18 arrived, however, the ship could not force its way to shore through the newly forming sea ice. It tried twice more in early March, but the captain finally gave up and steamed north to New Zealand for a second winter. "Campbell has certainly had very bad luck throughout," Debenham commented in his diary for March 4.[52] He did not know the half of it. The Northern Party's summer of fieldwork gave way to a winter of survival.

Of course, Scott's Polar Party also did not return on schedule in the summer of 1912, forcing the expedition to stay in place for a second winter. Among the scientists, Taylor and Simpson went home on *Terra Nova* in March and Lillie always worked from aboard ship, but the others remained south with most of the officers and men. Debenham did not know of Priestley's finds at Terra Nova Bay until the Northern Party returned, but during the winter he received the geological specimens collected by a support party that had accompanied Scott as far as the summit of Beardmore Glacier. "These had mainly been collected in the scattered moraine under Cloud-maker," a towering mountain on Beardmore's northern flank, the support party's leader, Edward Atkinson, reported. "To [Debenham's] surprise and joy several fossils of plants and small marine animals were found in some of these."[53] Then Priestley appeared in November with yet more Paleozoic fossils, which further filled out the story of Antarctica's temperate past. *Glossopteris* fossil finds came next. Following the Northern Party's near-miraculous return, only the waiting now remained for news of the Polar Party's fate. Having missed the chance to climb Mount Erebus with Shackleton's men, Priestley took this opportunity to lead an excursion up the volcano. He collected rock specimens and photographed the crater while Debenham surveyed the cone. This final outing marked the end of the expedition's geological fieldwork.

By the fall of 1913, Taylor, Debenham, Priestley, and Wright had reassembled for graduate studies at Cambridge, where they reflected on what they had learned about Antarctic geology. "It is now fifteen years since the first landing was made on the mainland of South Victoria Land," Debenham wrote in 1913. "The tale must necessarily be

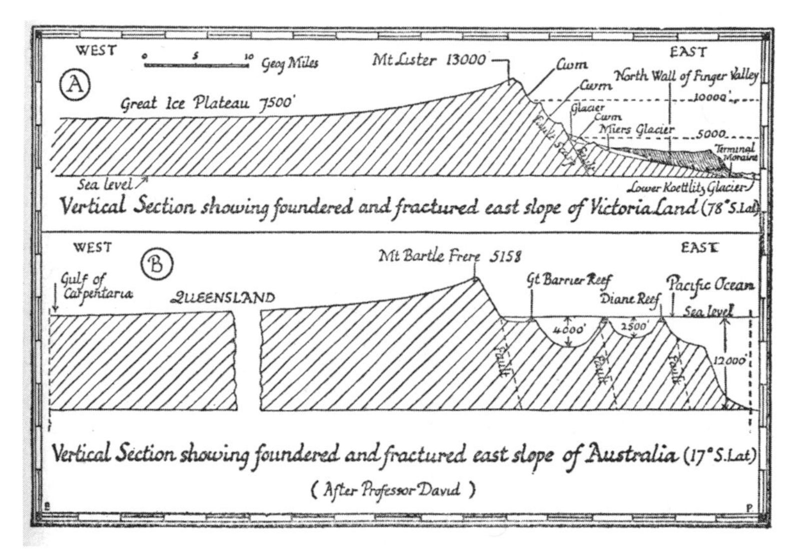

Comparison of the east coast of Victoria Land with the east coast of Queensland, Australia, by *Terra Nova* expedition geologist Griffith Taylor, from *Scott's Last Expedition* (New York, 1913).

incomplete, for the difficulties confronting geological investigation in those regions are naturally considerable, but enough has been done to warrant a preliminary interpretation of the known facts."[54] Clearly Antarctica was a continental landmass that once was connected to the other southern continents. This much was settled by the work of the three expeditions.

Using fossils finds and evidence of stratification, the geologists pieced together the region's past. As Ferrar had found, the gneiss, schist, and crystalline limestone came first, laid down underwater in a pre-Cambrian era. "The folding and heating of the rocks has since quite destroyed all evidence of the animal or vegetable life of that time," Debenham observed, "though numbers of small graphite particles, found in the crystalline limestone, may be the remnants of carbonaceous growth in the ancient coral reef." Following a long gap presumably caused by uplift of the land above sea level, a layer of limestone from the Beardmore region showed marine fossils from the Cambrian period. A profusion of igneous granites came next, the geologists induced, intruding during a time of uplift after deposit of the Cambrian limestone and before deposit of the late Paleozoic Beacon Sandstone, which reappeared at various sites along with shale, coal, or limestone. "There can be little doubt that these all represent deposits of approximately the same period under slightly different conditions," Debenham wrote. Next came the dark, igneous dolerite sill, "which stands out on cliff faces, produces pinnacled mountains, and generally dominates the topography." This sill eroded in diverse and dramatic ways to create Victoria Land's striking landscape. Fossils found during the *Terra Nova* expedition fleshed out the story. "The fish-remains from Mt. Suess," a report from that expedition stated, "point to a Devonian horizon, while the beds containing fragmentary leaves of *Glossopteris,* viewed in the light of the geological ranged of the genus in other parts of the world, may be as old as Upper Carboniferous or as late as the Rhaetic period."[55]

The region's distinctive modern features, the geologists surmised, arose on this foundation. Taylor compared Victoria Land to Queensland in his native Australia, with a sharp fault line running along the east coast. Breaking along this line probably in the early Cenozoic Era, a high plateau fronted by a dramatic coastal range uplifted in the west while a shallow sea basin sank in the east. "Simultaneously with, and probably as an effect of this faulting, there occurred a great outburst of volcanic energy along the line of breaking," Debenham suggested. "This outburst is now just dying out, only two volcanoes being still active, Mt. Erebus on Ross Island, and Sturge Island in the Belleny Group." The glaciers came after this uplift, Taylor reasoned, but before the volcanic activity subsided as the region cooled. "We know from the fossils that warmer conditions existed in Mesozoic times in Antarctica," he wrote.[56] The way Taylor and Wright read the glaciers, however, the cooling seemed over and a new warming under way. Building on the work of prior explorers, their study of Antarctic ice would conclude a final chapter in the science of the *Discovery, Nimrod,* and *Terra Nova* expeditions.

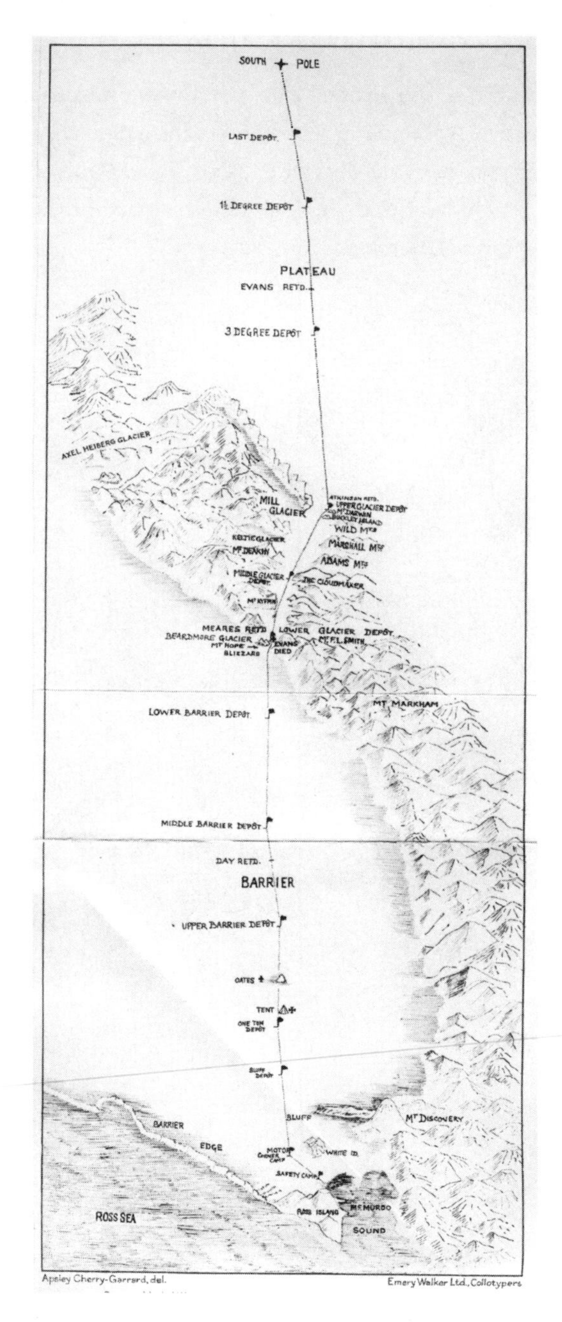

Polar journey of Robert Scott, Birdie Bowers, Edgar Evans, Titus Oates, and Edward Wilson (1911–12), from Apsley Cherry-Garrard's *Worst Journey in the World* (New York, 1922).

CHAPTER 8

The Meaning of Ice

ICE FEATURED PROMINENTLY IN THE THINKING OF all early Antarctic explorers. James Cook returned from his 1772–75 circumnavigation of the south polar region with the saga of a Great Southern Ocean blockaded by a field of sea ice. "This immense field," he wrote, "was composed of different kinds of ice; such as high hills; loose or broken pieces packed close together, and what, I think, Greenlandmen call field-ice. A float of this kind of ice lay to the S.E. of us, of such extent that I could see no end to it, from the mast head." From his 1839–43 voyage beyond that blockade, James Clark Ross added the image of a Great Ice Barrier extending hundreds of miles across the southernmost water approach to the pole: "It is impossible to conceive a more solid-looking mass of ice; not the smallest appearance of any rent or fissure could we discover throughout its whole extent, and the intensely bright sky beyond it but too plainly indicated the great distance to which it reached to the southward." The explorers who followed spoke of pack ice, land ice, ground ice, drift ice, icebergs, ice caps, ice falls, ice floes, floebergs, and glaciers. Clements Markham opened the *Discovery* expedition's scientific manual with a glossary of "Ice Nomenclature" that included over seventy such terms. In a 1900 address to the RGS, he listed "glaciation" as a primary field for "geographical discovery."[1]

In Britain, however, the lure of ice did not begin with Antarctic exploration. The drive to map the world's polar regions and bring them within the realm of scientific knowledge started in the north. The Franklin searches brought it to a head during the 1850s, but interest in such work was growing even before *Erebus* and *Terror* became locked in the Arctic ice, and it continued long after Leopold McClintock ascertained the lost expedition's fate. These episodes simply added to the allure.

These were heroic times in Britain, or at least a time in search of heroes, and polar firsts filled the bill. John Franklin's nephew Alfred, Lord Tennyson, Britain's poet laureate from 1850 to 1892, captured the spirit in verse. His epitaph for his uncle in Westminster Abbey read:

> Not here! the white North has thy bones; and thou
> Heroic sailor-soul,
> Art passing on thine happier voyage now
> Toward no earthly pole.[2]

The explorers themselves read Tennyson. Edward Wilson, a Christian ascetic and recovering consumptive, drew comfort on his polar journey from *In Memoriam*. It "makes me feel that if the end comes to me here or hereabout," he wrote from the base of Beardmore Glacier in 1911, "all will be as it is meant to be." When he spoke of humanity in this work, Tennyson seemed to speak of Wilson in particular: "Such splendid purpose in his eyes, / Who rolled the psalm to wintry skies." Wilson's sledge-mate and commander, the driven but self-doubting Scott, an introspective agnostic, surely noted the poem's portrayal of the human condition "as futile, then as frail!"[3]

The wanderlust depicted by Tennyson in *Ulysses* also spoke to the explorers and their conception of science: "To follow knowledge like a sinking star, / Beyond the utmost bound of human thought." And it added, "To sail beyond the sunset, and the baths / Of all the western stars, until I die." At the 1906 RGS ceremony honoring Roald Amundsen for completing the Northwest Passage, Nansen applied the final line of this poem to his Norse colleague: "To strive, to seek, to find, and not to yield." Many then saw a crystalline sea or field of ice as a fitting stage for humankind's heroic struggle against fate, nature, or futility. Markham extolled the *Discovery* expedition for giving navy officers like Scott "an opportunity to do battle with and to conquer the Antarctic ice, as our navy always conquers." In an age of Tennyson and Darwin, of "Nature, red in tooth and claw," science gave meaning to exploration, and glaciology became a purposeful part of polar expeditions.[4]

The very nature of ice posed questions. It was only in the late 1700s that chemists resolved the composition of water and determined the large amount of heat required to make ice melt. Differences in the types, colors, viscosities, and composition of ice puzzled physicists into the twentieth century. Sea ice and land ice raised separate issues, and distinguishing the former from floating land ice or compressed cakes of snow could baffle field researchers.

Ice sheets and glaciers compounded the problem. During the mid-nineteenth century geologists first recognized both the global extent of former ice sheets and role they played in shaping terrain. Although evidence of glacial scouring and transported debris persuaded them that vast ice sheets had once covered Europe and North America, they could not agree on how these seemingly solid bodies flowed like slow-moving liquid. Some thought that melting at the base caused the movement, others attributed it to gravity, and still others saw plasticity as a property of glacial ice. "Any mechanical theory of glaciers must be more or less imperfect which does not explain the remarkable veined or ribboned structure of the ice," an 1898 essay on the topic noted.[5] Greenland and the Alps offered some opportunity to study these questions, but nothing like the opening provided by Antarctica, where conditions resembled those of the ancient ice ages. Antarctic glaciers and ice sheets, researchers hoped, could shed light on those earlier ages and their associated climate changes.

Before his resignation, *Discovery*'s initial science director, J. W. Gregory, outlined an ambitious program of ice research for Scott's first expedition. "Members of the Expedition will have ample opportunities for the observation of Antarctic ice," he wrote, "but, that the observations may be scientifically useful, they must be systematic, and directed along certain definite lines." He proposed that the expedition begin by examining the icebergs and sea ice encountered on the way south. "The characteristic Antarctic icebergs are of the flat-topped variety, which in the Arctic region have been called floebergs from the belief that they were formed . . . by the freezing of layer after layer [of sea water] to the underside of the floe," he noted. The characteristic Arctic iceberg, in contrast, has a pinnacled top and calved from the sea edge of land ice. Which of these sources give rise to Antarctic bergs, he continued, "is still open. It can be tested by determining, in the case of Antarctic floebergs, the amount of salt in different layers." Ice floes themselves should be measured, he added, and their maximum thickness recorded.[6]

Turning to land ice, Gregory stressed the importance of studying both glaciers and the interior ice sheet. "That glaciers flow like rivers was discovered long ago," he noted, "but the cause and mode of this movement is still an open question. Many theories have been proposed in explanation of this, the most important of which are founded, directly or indirectly, on the study of the minute structure of glacier ice. . . . The size, shape and optical arrangement of the glacier grains should therefore be carefully studied." Gregory also urged that the explorers try to measure "the thickness of the Antarctic ice-sheet" on Polar Plateau. Estimates ranged from sixteen hundred feet to twenty miles, though anything over a few miles, he noted, was "not easy to reconcile with the known properties of ice."[7] He suspected that its greatest depth lay behind Victoria Land's coastal mountains and that the ice descended southward "slowly across the Pole to the shore of the Weddell Sea," where it discharged into the sea. Gregory called as well for detailed observations of glacial debris, deposits, scouring, and melting patterns. "Evidence as to whether the ice in Antarctic regions is now at its maximum extent, or whether it has anywhere retreated, should be carefully sought."[8]

The Great Ice Barrier raised an added set of questions. Nothing like it, Gregory noted, existed anywhere else on earth, and its origin "is still unsettled." Since Ross discovered it, many commentators had viewed the Ice Barrier as the level northward extension of a polar ice sheet, anchored on land in the south and pushed out over the Ross Sea, where it floated to a depth of more than a thousand feet. Others countered that it might either rest on the bottom of the southern Ross Sea basin or constitute the broad snout of a northward-flowing glacier. Returning from the *Southern Cross* expedition in 1900, Louis Bernacchi added his idea that the barrier was simply a long, narrow ice tongue bisecting the Ross Sea from west to east, the outflow of a Victoria Land glacier. Land might or might not exist farther south, Bernacchi told the RGS shortly before *Discovery* sailed, "but I do believe that there is an open sea between the southern side of the great ice-barrier and that land, if it exists."[9]

Gregory and others hoped the *Discovery* expedition would resolve these questions by sending parties across the Ice Barrier, recording its movement and looking for any detectable slope. In part to test Bernacchi's idea, Joseph Hooker, the last surviving member of Ross's expedition, proposed that the ship carry a tethered hot-air balloon so that the explorers could look across the Ice Barrier from its seaward edge. Like Ross, Hooker thought that ice extended to the South Pole. "Of course, the great interest of the voyage and observations is that great ice-barrier," he told the RGS. From the balloon—inflated only once at an inlet in the Ice Barrier—Scott, Shackleton, and three other Antarctic aeronauts saw nothing to the south but what Shackleton described as "league upon league of whiteness." The Ice Barrier was a true ice sheet or shelf flowing from the south, not an ice tongue extending from the west, but it still might originate on land short of the pole. In a poem composed after his balloon ride, Shackleton asked of the Ice Barrier, "Shall we learn that you come from the mountains? Shall we call you a frozen sea?" Either explanation remained plausible.[10]

From these ice studies, expedition organizers hoped to learn about more than local conditions. "The study of Antarctic ice is desirable," Gregory wrote, "from the evidence it may yield as to the conditions of the period when some form of ice agent deposited a vast sheet of clay, sand and gravel over much of North-western Europe, including most of the British Isles." The deposits could have come from icebergs calved from a vast network of ancient glaciers or from a massive ice sheet covering the entire region. "No evidence in support of the latter hypothesis is given by Arctic ice," Gregory noted, "but the great Antarctic ice barrier may conceivably be doing what some geologists assure us that the Scandinavian ice-sheet did in 'the Great Ice Age,' and what other geologists tell us is impossible."[11] If so, the lessons learned from Antarctic ice could carry global significance. The Duke of Argyll, a leading British advocate of the theory of a geologically significant former ice age, made this point in promoting the *Discovery* expedition. The most discussed and difficult geological questions of the day, he asserted, related to how moving ice sheets affect land. "These questions, and a hundred others, have to be solved by Antarctic discovery; and until they are solved we cannot argue with security on the geological history of our own temperate regions."[12]

Gregory's resignation hobbled the expedition's efforts to study Antarctic ice, much as it hampered work in geology. As a field geologist, Gregory had cut his teeth studying the impact of glaciers on the American Rockies. He pursued similar research in the Alps and participated in a major Arctic expedition, where he studied both sea and land ice in the Spitzbergen region. "His ice experience," Gregory's chief supporter on the Royal Society's Antarctic Committee declared, was "of the highest importance."[13] In replacing him, Scott made no effort to secure someone with expertise in glaciology. Responsibility for the discipline fell on the expedition's substitute geologist, Hartley Ferrar, who was barely qualified in his own field and had never worked in polar or alpine regions. Fortunately, as a last-minute fill-in to make magnetic observations, Scott signed on Bernacchi, who had interest in the Ice Barrier from his experience on the *Southern Cross* expedition and would champion its study.

On the voyage through the Southern Ocean and into the Ross Sea, expedition members made regular observations of icebergs. Charles Royds spotted the first one on January 2, 1901, Wilson noted. "Soon Skelton sighted another and I a third and then there were never less than 4 or 6 in sight at once." They counted, measured, and photographed them. Wilson sketched many. "The nature and origin of the southern iceberg have always been a subject of some mystery," Scott wrote. Unlike those in the North Atlantic, he observed, "they have all a flat top and wall sides, and appear to have broken quietly away from some huge sheet of ice." Some were much larger than any in the north. "The largest berg we saw was aground off King Edward's Land," Scott wrote in some awe, "and we estimated it as about seven miles long and 200 feet high." From their size, shape, location, and movement, the explorers concluded that Antarctic icebergs detached either from the Great Ice Barrier, which Shackleton hailed as the "mother of mighty icebergs, those Kings of the Southern Seas," or from one of the region's floating glacial ice tongues. Given these sources, Scott commented, "I see no reason why their length should be limited." Unlike Arctic floebergs, they were composed of salt-free ice that clearly originated as snow.[14]

Two days after first sighting icebergs, the explorers entered the sea-ice pack. The expedition's second-in-command and only member with Arctic experience, Albert Armitage, wrote, "To one accustomed to the ice-pack of the Northern seas, this Southern ice presented a most extraordinarily level appearance. It seemed to me as though large fields of ice had been formed in protected places, and then broken up." Scott added, "The nature of the pack seemed to change every few hours; sometimes the floes were more easily pushed aside and broken, at others they flew apart at the first shock; at times the prow entered deep in a floe before the first crack appeared, at others it seemed to make little impression."[15]

The study of sea ice continued after the explorers left the pack. "During our long stay in our winter quarters," Soctt wrote, "we were able to observe to some extent the breaking-up and clearing of the Ross Sea, which goes to form this line of pack." Based on data collected at field sites, Ferrar computed that salt water in the Ross Sea froze to a depth of eight feet during winter, broke up by summer, and floated north to create the surrounding pack. "The upper two inches consist of plates, a quarter of an inch across and a sixteenth of an inch thick, which lie horizontally, and only gradually do these give place to the sheaves of vertical fibers which make up the greater mass of the ice," he reported. "The salinity seems to depend more upon the rate of freezing than upon the depth or distance from the surface." Ferrar found sea ice much less salty than sea water. Further, once the sea froze over, snow accumulated on top and pushed down the ice, which melted from the bottom so that at some sites all the original sea ice was totally dissolved by spring, leaving only ice from the snow.[16]

After exiting the pack and crossing the Ross Sea, *Discovery* sailed back and forth for nearly a thousand miles along the Ice Barrier's face before entering winter quarters. Observations made during this passage established that, at least at its sea edge, the Ice Barrier floated and was calving rapidly. Plotting *Discovery*'s route along the barrier's front against Ross's path in 1841, Scott found that the sea edge had retreated by as much as thirty miles in sixty years. "We had sailed continuously over ground which in [Ross's] day had been covered with a solid ice-sheet," he noted. Sounding lines found sufficient depth to float ice of the barrier's height and drew up pebbles from the seabed, suggesting that the Ice Barrier's underside dropped rock debris scoured from the Antarctic mainland. "The part that has broken away must therefore have been water-borne, and this at least shows the possibility of the ice-sheet being afloat for an almost indefinite distance to the south," Scott reasoned. Further, he observed, when *Discovery* tied up to the barrier, "although we had evidence of considerable tidal movement, the ice rose and fell with the ship."[17] The explorers later found added evidence that the Ice Barrier floated by studying its movement at tide cracks where it abutted land in the west.

After *Discovery* anchored at its winter quarters, where it remained frozen in place from February 1902 to February 1904, none of the explorers made ice a primary subject for research. Indeed, after one summer passed without McMurdo Sound breaking up, their main concern with sea ice was freeing the ship from it. Nevertheless, Scott's Southern Sledge Journey in the summer of 1902–3 demonstrated that the Ice Barrier extended for at least three hundred miles southward, with no end in sight. "It was a surprise to everyone, and not least to ourselves, to find that our long journey to the south was made without a rise of level," Scott noted. "I do not see that there can be any reasonable cause to doubt that the Great Barrier ice-sheet is afloat at least as far south as we travelled." That same summer, Armitage's Western Sledge Journey gave humans their first glimpse of the earth's largest ice sheet—the Polar Plateau. "There was nothing but an undulating white surface," Armitage wrote, "we were, in fact, on the summit of the ice-cap in that portion of Victoria Land, at a distance of 101 statute miles from the coastline."[18] Throughout the expedition, while focusing his efforts on geology, Ferrar periodically checked the depth, structure, and salinity of sea ice at sites near winter quarters. The ship's confinement brought an unexpected third season of sledging, giving the explorers a chance to conduct three additional journeys that shed light on Antarctic ice.

* * *

Bernacchi talked Scott into the first of these ventures. Perhaps still nursing his idea concerning open water in the south, Bernacchi argued that despite its first year's efforts, the expedition had learned little about the interior of the Ice Barrier. Scott's Southern Party had merely found that it extended level for some three hundred miles on its western edge. Bernacchi proposed sledging diagonally across the Ice Barrier in a southeasterly direction from Ross Island. Scott tapped Royds to lead the party, which included Bernacchi and four from the ship's company. Man-hauling two five-hundred-pound sledges, they left on November 10, 1903, with instructions to determine whether the Ice Barrier continued level toward the east and to return in time to help free *Discovery* from the ice for the voyage home. "It was a short journey, as it only occupied thirty days, and for those who took part in it it could not be otherwise than monotonous and dull," Scott wrote, "yet it deserves to rank very high in our sledging efforts, for every detail was carried out in a most thoroughly efficient manner."[19] Bernacchi used the occasion to take a regular series of highly accurate magnetic readings, far from disturbances caused by land or the ship's metal fittings.

Sledging southeast for 178 miles, or about one-third of the way across the Ice Barrier, Royds's party found nothing except an icy, undulating surface with no noticeable net rise. Scott called this "a negative but highly important result" and recognized that it came at a cost. "The party went on a very short food allowance," he wrote, "and day after day found themselves marching over the same unutterably wearisome plain." Bernacchi added, "This dragging was very heavy on account of the bad surface, and the temperature was generally below zero, with a strong wind and drifting snow." Everyone felt real hunger. "'Food, food, food' is what one thinks about," Bernacchi noted. "We, of course, saw no new land, nothing but the barren, level surface of the barrier." Years later, he still remembered the "ghastly business" of relieving himself on a featureless plain in high winds and blowing snow. "There are no facilities within the tiny tent; latrines are impossible," he observed, "so there remains the open snow spaces." After loosening layers of clothing, he explained, "the rest is a matter of speed and dexterity, but invariably the nether garments are filled instantly with masses of surface-drifting snow." This snow caused extreme discomfort as it melted. "Low temperatures, painful though they may be, are not serious, but there have been cases of quite unpleasant frostbite."[20]

Even as Royds's party struggled across the Ice Barrier to the southeast, a sledging party led by Michael Barne made a dramatic discovery in the south. Scott had sent Barne's party south in October 1903 to explore the junction between the Ice Barrier and the South Victoria Land coast. "Ill fortune dogged this party from the start," Scott reported. "They were hampered with continual gales from the south, and again and again had to spend long days in their tents." When Barne and his men tried to cross over the divide separating the Ice Barrier from the coast to study the land, they were blocked by the same sort of deep crevasses and steep hummocks that had stopped Scott from making a similar traverse during his Southern Sledge Journey. "From their observations," Scott concluded, "it is doubtful whether a sledge party could ever cross it unless they were prepared to spend many weeks in the attempt." With no such time, Barne's party turned back in late November having traveled about 150 miles from winter quarters, with little more than a detailed map of the coastline to show for the effort. On the return trip, however, Barne noticed that a well-marked supply depot placed on the Ice Barrier over thirteen months earlier had moved 608 yards seaward. "Almost accidentally we obtained a very good indication of the movement of the Great Barrier ice-sheet," Scott reported.[21]

The season's signature event—Scott's Western Sledge Journey—began and ended in Victoria Land's coastal mountains, but the middle was epic ice. In November 1903, after retracing Amitage's route up nine thousand feet in one hundred miles to the Polar Plateau with five others and, for most of the way, Ferrar's three-person geology team, Scott set out across it with two navy bluejackets, Edgar Evans and William Lashly, whom he viewed as the ablest sledgers in the group. He depicted Evans as "a man of Herculean strength, very long in the arm and with splendidly developed muscles." Scott, who was a heavy smoker and felt its effects, marveled at Lashly's physical discipline and massive chest. The thirty-six-year-old navy stoker, he noted, "had been a teetotaler and non-smoker all his life, and was never in anything but the hardest condition."[22] No one had ventured onto the Polar Plateau before. With these men, Scott tried to cross it: the highest, coldest, driest, and largest ice sheet on earth. Perhaps more than anything else in Antarctica, this utterly unique mass of ice, which holds over half of the earth's fresh water, aroused the curiosity of glaciologists. Scott's account introduced them to it.

Scanning the ice westward from the summit of the coastal mountains, Scott had some notion of what to expect. "I do not think it would be possible to conceive a more cheerless prospect than that which faced us . . . on this lofty, desolate plateau," he wrote. As he had done in the Southern Sledge Journey, he again faced crossing a seemingly endless plain of ice, this time without a dramatic coastline to follow. "Yet before us lay the unknown. What fascination lies in that word! Could anyone wonder that we determined to push on?" Still, conditions proved worse than he expected. The weather was colder than any sledge party had yet endured for such a long period. "Regularly each night, when the sun was low in the south, the temperature fell to −40° or below, whilst during the marching hours it rarely rose much above −25°," Scott noted. "If the thermometer can fall to −40° in the height of summer, one can imagine that the darker months produce a terrible extremity of cold."[23] He had expected the high altitude and cold temperature to bring clear skies, but a heavy overcast sometimes forced the party to stop midday. "We could not see a foot in front of us," Lashly complained at one point. "When you get a gray day up there you can see nothing," Scott added. "From under your feet to the zenith there is nothing in sight." Further, near the end of the return trip, an icy snowfall hampered sledging.[24]

A fierce westerly wind made matters worse. "It has cut us to pieces. We all have deep cracks in our nostrils and cheeks, and our lips are broken and raw; our fingers are also getting in a shocking state," Scott wrote after one week on the plateau. "We can do nothing for this as long as we have to face this horrid wind." He had determined from the outset to march due west until the end of November and then turn around: two weeks out and two weeks back. The stopping place was purely artificial—there was nothing to see at any point except more of the same—yet Scott held to it. The men carried just enough food and fuel for the planned trip, a tactic that proved problematic near the end when overcast skies and snow slowed their return. Both the fuel and Scott's tobacco ran out, causing irritation all around. From his experience in the howling wind and by observing drift patterns in the surface, Scott concluded, "The wind blows from west to east across this plateau throughout the winter, and often with great violence."[25]

Beyond reports on wind and weather, Scott brought back a vivid picture of the plateau's surface. Using a fluidless barometer, he determined that the net altitude remained nearly constant at some nine thousand feet across the two hundred miles of ice sheet that his party crossed. The surface itself featured broad undulations, averaging three to five miles from rise to rise, overlaid by a pattern of wavelike sastrugi running parallel to the prevailing wind. "The summits and eastern faces of undulations were quite smooth with a very curious scaly condition of surface, whilst the hollows and the western faces were deeply furrowed with the wind," Scott reported. "For long stretches we travelled over smooth glazed snow, and for others almost equally long we had to thread our way amongst a confused heap of sharp waves. I have rarely, if ever, seen higher or more formidable *sastrugi*." He compared the journey to sailing on "a small boat at sea: at one moment appearing to stand still to climb some wave and at the next diving down into a hollow. It was distressing work, but we stuck to it, though not without frequent capsizes." Otherwise, the surface remained constant, with none of the crevasses and hummocks that marked glaciers or sea ice. "Could anything be more terrible than this silent, wind-swept immensity?" Scott asked in his diary.[26]

The gravest threat to life or limb came as the party left the plateau. In rising winds, the men lost control of the sledge on the first icefall descending to the glacier and were jerked forward by their harnesses. "We all three lay sprawling on our backs and flying downward with an ever increasing velocity," Scott reported. "At length we gave a huge leap into the air, and . . . came down with tremendous force on a gradual incline of rough, hard, wind-swept snow." Fifteen minutes after resuming their march and still tethered to the sledge, Evans and Scott stepped into a crevasse. "Personally I remember absolutely nothing," Scott wrote, "until I found myself dangling at the end of my trace with blue walls on either side and a very horrid-looking gulf below." Evans dangled nearby. The sledge tottered on the edge, twelve feet above them, with only Lashly's extreme effort keeping it from taking everyone down to certain death. "It is some time since I swarmed a rope, and to have to do so in thick clothing and heavy crampons and with frost-bitten fingers seemed to me in the nature of the impossible," Scott noted, but he managed. After securing the sledge, Scott and Lashly pulled up Evans. "We all agreed that yesterday was the most adventurous day in our lives," Scott wrote in his diary a day later, "and none of us want to have another like it."[27]

Descending from the Polar Plateau, Scott's party discovered the Dry Valley, which clearly showed one glacier in full retreat. Having seen the vast expanse of ice on the plateau, and now noting signs of glacial scouring and debris high on the valley walls above the descending glacier, Scott became convinced that the polar ice sheet was shrinking. "At a comparatively recent period," Scott reported to the RGS, "the whole glaciation of the region was vastly more extensive than it is today." Similarly Ferrar, in his report on Antarctic geology, repeatedly referred to "the recession of ice," "diminution of ice," and "a former greater ice supply" as characteristic of the region.[28]

The expedition's finding that the Ice Barrier's edge had retreated since Ross's day supported this hypothesis. "It is evident," Scott wrote, "that when the Southern glaciation was at a maximum, when the glacier valleys were filled to overflowing, and when the great reservoir of the interior stood perhaps 400 or 500 feet above its present level and was pouring vast masses of ice into the Ross Sea, the Great Barrier was a very different formation from what it is at present." He saw it as a remnant of an ice sheet that once filled the Ross Sea basin and extended into the Southern Ocean: "There are abundant evidences of its great enlargement; granite boulders were found on Cape Royds and high on every volcanic island in our neighbourhood; on the slopes of Terror, Dr. Wilson found morainic terraces 800 feet above the present surface of ice; Mr. Ferrar showed that nearly the whole of the Cape Armitage Peninsula was once submerged; and, in fact, on all sides of us and everywhere were signs of the vastly greater extent of the ancient ice-sheet."[29]

To Scott, present-day Antarctica and its current ice sheet resembled prehistoric northern Europe and its former ice sheet. Already the Antarctic ice pack appeared to dissipate entirely each summer. The warming might continue. "It is strange to think," he concluded, "that there may be a season in the year when the enterprising tourist steamer may show its passengers the lofty smoke-capped form of Mount Erebus as easily as it now does the fine scenery of Spitzbergen."[30]

The *Nimrod* expedition picked up where *Discovery* left off in examining the striking retreat of Antarctic ice. Primed by the comments of Scott and Ferrar, Shackleton's geologists, Edgeworth David, Douglas Mawson, and Raymond Priestley, found evidence of diminished ice everywhere they looked. It became their all-purpose explanation for such disparate observations as dry valleys, dead glaciers, raised beaches, stranded moraines, isolated ice fields, granite erratics on volcanic mountains, glacial scouring high on rounded valley walls, and Victoria Land's broad coastal shelf. Despite enduring extreme cold and harsh blizzards, these geologists repeatedly found signs of a shrinking ice sheet on their northern and western sledge journeys and ascent of Mount Erebus.

"Certainly the Antarctic ice sheet is decreasing rapidly," David and Priestley stated in the expedition's official report, "and this ice shrinkage has been general." Describing their own observations, they wrote in 1910, "The evidence collected by the Northern and Western parties of formerly far more intense glacial action than at present extends from Mt. Nansen to the Ferrar Glacier, a distance of about 200 geographical miles." Extending their account to cover the entire Ross Sea basin, they added that expedition members found erratics high on Mount Bird, sixty miles north of the Ice Barrier's front; on Mount Hope, near Beardmore Glacier far to the south; and on Mount Erebus in the east. In a 1909 list of his expedition's contributions to geology, Shackleton named first the finding that "throughout the whole of the region of Antarctica examined by us for 16° of latitude there is evidence of a recent great diminution in the glaciation." He estimated that, based on the altitude of moraines on Mount Erebus and the depth of McMurdo Sound, the Ice Barrier "at its maximum development must have had a thickness of not less than 2800 feet" at points where it no longer existed, and noted its continued retreat through calving.[31]

Using the evidence they collected, David and Priestley attempted to determine the extent of maximum recent glaciation and map the Ice Barrier's former limits. They calculated that the glaciers running

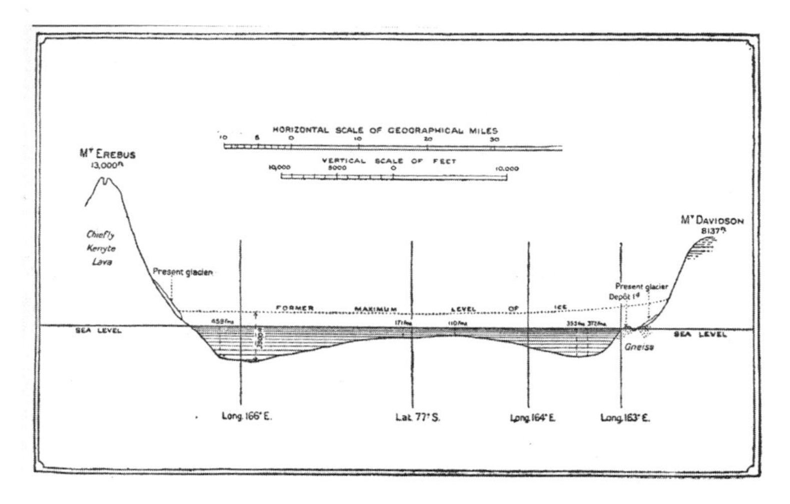

Sketch of former glaciations in the Ross Sea basin by *Nimrod* geologists Edgeworth David and Raymond Priestley, from Ernest Shackleton's *Heart of the Antarctic* (Philadelphia, 1909).

from the Polar Plateau to the Ice Barrier or Ross Sea—Beardmore, Ferrar, Mackay, Reeves, and others—had stood two to four thousand feet higher than at present, and the plateau itself had probably emptied out significantly. Now that sledge parties had crossed long stretches of the plateau at two additional points, in the north to the magnetic pole and in the south toward the geographic pole, David and Priestley recognized that it rose from about seven thousand feet above sea level in the north to more than ten thousand feet in the south, with Scott's earlier finding of nine thousand feet roughly in the middle. Given the height of the stratified rock layers exposed in the Victoria Land mountains, David and Priestley placed the depth of the polar ice sheet at less than three thousand feet—only a fraction of later estimates. The sheet "moves slowly coastward," they wrote, "either down to the Great Ice Barrier or to the western shore of McMurdo Sound and the Ross Sea." The barrier itself, they estimated, "attained a maximum thickness formerly of nearly 4000 ft. in parts of McMurdo

Sound from which it has now entirely retreated" and once extended at least seventy-five miles north of its present edge.[32]

A chance discovery led to a revised understanding of the Ice Barrier. Scott had depicted the Ice Barrier as a shelf of floating ice with the buoyancy of a glacial berg. On a supply journey across the barrier, Ernest Joyce, who had also served on the *Discovery* expedition, recognized the flag of a depot laid by Scott on October 1, 1902—the same depot whose movement Barne had measured in 1903. "There was a bamboo pole about eight feet high projecting from the snow, with a tattered flag," Shackleton wrote. "The guys to which the pole was attached were completely buried under the snow." Joyce found that the depot had moved 9,600 feet northeast of its initial placement and was beneath ninety-eight inches of compressed snow. From this, the explorers calculated that the Ice Barrier traveled seaward at a speed of roughly one mile every three years and gained about one foot of packed snow annually. At this rate, a point on the barrier would add nine hundred feet of compact snow during its three-hundred-mile journey from the mountains to the sea. This must press down the glacial ice and cause it to erode from below. "Probably ere it reaches the Barrier edge the ice may be entirely thawed away by the action of the sea-water, leaving only the floating snow-sheet," *Nimrod* naturalist James Murray told a Scottish audience. "This theory suggests that the Barrier is an accumulation of snow" rather than glacial ice, Shackleton explained to the RGS. As such, it should float higher than previously thought.[33]

To test their theory, the explorers examined three barrier bergs near winter quarters and found them composed of dense snow and grounded in shallow water. "There can, we think, now be little doubt that a great proportion, in some cases the whole, of the material of typical Antarctic bergs is formed of consolidated snow rather than ice," David and Priestley wrote, and extrapolated from this to conclude that "there was every appearance of the Barrier being formed of numbers of superimposed layers of snow."[34] On his Northern Sledge Journey, David also crossed two ice tongues extending from Victoria Land glaciers into the Ross Sea. Both were at least partly afloat, he surmised, and composed largely of glacial ice and packed snow. In addition, during the winter, Mawson, Priestley, and Murray studied the formation and structure of ice in the lakes around Cape Royds, which involved sinking shafts through the ice to the lakebeds, recording the shape of crystals, and monitoring the temperature and salinity at various layers.

Scott returned to the Antarctic in 1910, just as debate over the Ice Barrier's history and composition gained scientific currency and public attention. Though on a mission to reach the pole, he would not ignore research questions—especially questions on topics he had already addressed, such as the barrier. "The main object of our expedition was to reach the South Pole," Scott's second-in-command, Edward R. G. R. "Teddy" Evans, explained, "but the attainment of the Pole was far from being the only object in view, for Scott intended to extend his former discoveries and bring back a rich harvest of scientific results. Certainly no expedition ever left our shores with a more ambitious scientific programme."[35] Priestley, back in the field after only a year, also had an established interest in Antarctic glaciology in general and the Ice Barrier in particular.

Scott dispatched Priestley and a small party by ship with instructions to winter on King Edward VII Land near its junction with the Ice Barrier, explore southward, and then return over the barrier to the main base during the next summer. Led by the expedition's third-in-command, Victor Campbell, the party included medical officer Murray Levick and three sailors, George Abbott, Frank Browning, and Harry Dickason. Barred from reaching King Edward VII Land by pack ice and deterred by Amundsen's presence from staying on the barrier at the nearby Bay of Whales, the so-called Northern Party spent its first winter at Cape Adare and was redeployed for its next summer by ship to Terra Nova Bay, on the Ross Sea coast 250 miles south of Cape Adare.

At both sites, the party's main mission was to make scientific observations. The effort again showed Scott's willingness to devote precious resources and some of his best men to research activities. Unlike Amundsen's, with its single-minded pursuit of the pole, Scott's expedition had mixed goals.

Terra Nova Bay was a fine place for a month of geological and glaciological research in summer, but in winter, after the ship was blocked by the pack from retrieving the party, it became an awful place to live. The Northern Party recognized its fate in late February 1912 after waiting nearly two weeks at Hells Gate, the aptly named drop-off and pick-up location. "Most of the time while we were waiting," Campbell noted, "the wind blew with uninterrupted violence and the tents suffered considerably." Hells Gate stood at the sea edge of a frozen basin fed by steep mountain glaciers and separated from the bay by low coastal islands or moraines. During virtually the entire winter, fierce winds from the Polar Plateau swept down the glaciers and across the basin to the sea. Gusts tossed the men to the ice or into boulders. At times they had to crawl. "There was nothing that strained our patience so near to the breaking-point as did this wind," Priestley wrote. The men could not survive in tents and had brought only enough food and fuel for their planned short summer stay. Further, the sun circled ever lower in the horizon and soon would disappear for over two months, plunging temperatures far below zero—though how far below they would never know, because they had lost their last thermometer. "It was evident that three things were absolutely necessary," Priestley wrote. "We must have light, shelter, and hot food."[36]

A mile from their Hells Gate depot, Campbell and Priestley found a large drift of packed snow on a boulder-strewn coastal ridge they named Inexpressible Island. "We first sank a trench 3 feet by 4 feet to a depth of 6 feet," Priestley wrote, "and then from the side of this we picked out a large cave toward the thickest part of the drift." When finished the cave—twelve feet wide by nine feet long and not quite tall enough for a man to stand—became the party's winter quarters. "Our roof is of hard snow about 3 feet thick, while the walls and floor are of ice," Campbell added. Roofed with seal skin and snow, the trench served as a storage vestibule with a hatch for coming and going. The men heated and cooked with a jury-rigged tin stove that burned seal blubber and caked everything inside the cave with greasy black soot, including the men. "Roof, walls, sleeping-bags, cooking utensils, and food-boxes became blacker and blacker," Priestley wrote, "and the smoke from the blubber-stoves during the day nearly drove us mad." They ran a risk of asphyxiation whenever the chimney and hatch became stopped with snow. To prevent the ceiling from melting, they kept the inside temperature below freezing and, when not doing chores, spent day and night in their threadbare sleeping bags.[37]

With shelter found, food became their preoccupation. Campbell tapped Priestley to ration the stores. From the scant remaining supplies, Priestley set aside the barest minimum for the anticipated spring sledge journey back to Ross Island and divided the rest by the time ahead. This allocation left each person with only a biscuit and mug of weak cocoa or tea per day, twelve lumps of sugar and two ounces of chocolate every week, and twenty-five raisins monthly. Priestley saved the pemmican for sledging. Beyond these rations, the party lived on what it killed, which meant every seal and penguin that wandered its way that winter. There were never enough. "We were all miserably hungry," Campbell wrote.[38]

Seal became the staple, with the meat and organs cooked into stew or "hoosh" and the blubber fried, eaten raw, or used for fuel. Everyone found seal meat so insipidly bland that the cooks made the hoosh with sea water for added salt and raided the medicine kit for flavoring, with mixed results. To the general dismay, hoosh stewed with a mustard plaster tasted more of linseed than mustard, but, Priestley noted, "even this was a change from undiluted seal."[39] Biscuits ended in August, sugar and chocolate in September; but seal hoosh continued, even after a tainted batch caused food poisoning. Dysentery became endemic, Campbell observed, "which is most inconvenient in this weather when to be out in the wind for a few seconds means a frostbite." The meat-only diet also caused a loss of bladder control. "Some of us," he wrote, "have been wetting our clothes in our sleep. One of us even while he was awake before he could get out of the door—and these are the clothes we have to live in for about 8 months."[40] The diet affected Browning worst: he became so weak that he could not help with sledging in the spring. Everyone had sore joints from limited food and exercise.

Light for the windowless cave came from burning wicks in tins filled with oil extracted from seal blubber. The light made it possible to work, move about, and read. The men had three books and two magazines among them, which were read aloud for after-supper entertainment. They also conducted songfests many evenings. Levick occasionally lectured on anatomy, and Priestley read his diary to the others. After supper on Sundays, Campbell recited from scripture, and Wesleyan hymns replaced sailors' songs. Most of all, the men planned their escape. All sought to maintain civility by complaining only about their common plight and feared for their sanity should they not get out in the spring. "No cell prisoners ever had such discomforts," Teddy Evans wrote in his diary when *Terra Nova* stopped at the site a year later.[41]

Whenever the wind died down during the long winter darkness, members of the party ventured out for food. If they found a seal, they butchered it on the spot and lugged back the joints of meat and slabs of blubber. Everyone's clothes became tattered rags, so soaked with grease and smeared with soot that stitches would not hold. "We have been in the same clothes for 9 months, carrying, cooking, and handling blubber," Campbell wrote in late September. "In fact we are saturated to the skin with blubber." Washing was impossible. For Priestley, boots posed an even greater problem than clothes. "If they had not always been frozen," he observed, "they would have dropped to pieces." The party saved its reserve clothing and footwear for spring sledging. "Throughout the whole of the winter the winds were so bad that except for passing observations science was definitely impossible," Priestley lamented. "We seldom even found it possible to take any walks for the sake of exercise."[42]

Somehow, the party survived the winter and, in spring, hauled two sledges over two deeply creviced ice tongues and two hundred miles of often disturbed sea ice to Cape Evans. With Browning unable to pull and others occasionally incapacitated, the trip took longer than expected—thirty-eight days in total. Until they reached the depot at Butter Point, the men marched on half-rations and whatever they could kill—often relaying their two sledges one at a time, which tripled the distance. "We were frequently floundering for several yards together up to the sockets of our thighs in snow," Priestley commented on the early spring sledging conditions. "To add to our difficulties, the light was abominable and snowsquall succeeded snowsquall with slight intervals between."[43] The sledges regularly capsized on slopes, and one finally broke down altogether. Nevertheless, Priestley gathered rock specimens along the way and stopped to pick up collections left by David's Northern Party in 1909, which the *Nimrod* expedition had failed to retrieve. The party returned with a tale of endurance in ice unsurpassed in the annals of British polar exploration—all done in the name of science.

Even as the Northern Party endured two winters forestalled or beset by ice, the expedition's other scientists and officers pursued research involving glaciers, snow crystals, and sea ice. In a place where the snow in summer extended down to the sea and most land was covered by ice, this choice of subjects was logical.

From the outset, senior geologist Griffith Taylor accepted David's theories of decreased Antarctic glaciation and of ongoing climate change. In an article for the expedition's newspaper published during the first winter, he explained to his colleagues that the Ice Barrier once overrode "the shores of MacMurdo Sound" and its weight pushed up the lava that enlarged Mount Erebus and gave rise to Cape Evans. He wrote of a future time when "still warmer conditions supervene" and predicted that "a topography like that of the Himalayas" would emerge from under the Antarctic ice sheet. A eugenicist who would devote much of his subsequent academic career to promoting hierarchical theories of racial anthropology, Taylor speculated that "in the moraine-fed troughs of the Ferrar and Dry Valleys will dwell a white race, depending partly on the fertile glacial soil, but chiefly on tourists from effete centres of civilization." Those tourists, he quipped, "will proceed in the comfortable steamers of the Antarctic Exploitation Company to the chalets of Beardmore. Here start the summer motor trips to the South Pole."[44]

During his fourteen months in Antarctica, Taylor led the two Western Sledge Journeys that explored a hundred-mile stretch of the Victoria Land coast opposite Ross Island. While Frank Debenham collected rocks and fossils, Taylor studied glaciers. The region included four outlet glaciers from the Polar Plateau: Koettlitz Glacier in the south; Ferrar and Taylor in the middle; and Mackay Glacier with its long ice tongue in the north. "The four were diverse enough to embody almost the whole cycle of glacial erosion within their domain," Taylor noted. He mapped them, measured their movement, and examined their impact on terrain. Of the four rivers of ice, Taylor found Mackay's terminal ice tongue moving the fastest—roughly a yard per day. "She is fairly galloping to sea," he noted in his diary. Each glacier had carved out mountain valleys and coastal harbors in the past, Taylor concluded, and was once much larger than now. This suggested an earlier period of more snow and ice. "Later the snowfall diminishes and the erosive power decreases," he wrote. "The Mackay with a few 'hanging' glaciers, the Ferrar with a preponderating number of tributaries hanging on the slopes of the main trough, are examples of the earlier stages in this decline. The Koettlitz with its tributaries 5 miles back from the main glacier and the Taylor Glacier with its extraordinary ice-free outlet trough 25 miles long are later stages in the retrocession of the ice mantle."[45]

These findings had meaning beyond Antarctica. In Victoria Land, Taylor asserted, scientists could witness stages in the glacial action that once shaped the landscape of central Europe and other mountainous regions. "There can be no more valuable branch of geology," he wrote, "than one which tries to chronicle the actions which have made the Alpine countries of the world so different from the more normal regions." On the Western Sledge Journeys "I was to find that many of the features in Antarctica reproduced, in the present, the past history of the Swiss scenery." Priestley and Charles Wright later added, "Modern theories of ice action can be tested only in the one region where ice sheets of continental extent still maintain almost their fullest development. It is in the Antarctic alone that all the main types of land ice are met with today."[46]

Wright accompanied Taylor on the first Western Journey, which departed in January 1911. A physicist, Wright studied the movement of glaciers with Taylor; on his own, he examined the formation and structure of ice and snow crystals. Wright carried a large camera to photograph the crystals and a bulky polariscope to examine sheets of ice in polarized light. Further, he cut into glaciers to examine the interior structure. "Magnificent crystals seen in pits reaching through the 3 in. snow to the clear(ish) ice," he reported on the party's first day. "To photo crystals tomorrow."[47]

A hearty Canadian who showed his toughness by walking from

Cambridge to London to apply for a position on the expedition, Wright loved ice. In mid-March, one day after returning from the Western Sledge Journey and eager to get onto the Ice Barrier, Wright volunteered to help Scott on the final depot-laying trip before winter. The journey proved exceedingly cold and offered few chances for research, but Wright's enthusiasm impressed Scott, who wrote that the Canadian "has taken to sledging like a duck to water." In June, when Scott invited him to serve as the only trained scientist on the Polar Journey, Wright exclaimed in his diary, "Gott sei dank!" or "Thanks God!"[48]

During the first winter, before the Polar Journey, Wright used his time at Cape Evans to extend his ice research. He observed the formation of sea ice, experimented with creating ice crystals at various temperatures, measured the amount of air included in snow and ice, studied the ablation of glaciers, and examined ice crystals formed by fog and humidity. "Walked down southeast a couple miles along the sea ice to some ice caves," he wrote in one midwinter diary entry. "Found some magnificent crystals (bell shaped) in one of the crevasses. They grow in huge clusters almost the size of one's head."[49] Depending on temperature, he found, Antarctic snow fell in a range of forms, from delicate six-rayed stars to hexagonal plates and granular balls, and by changing its crystal structure, it condensed first into white névé and then into blue ice. "During the summer one can see the whole transformation taking place before one's eyes in the course of a few days," he added later. Wright also examined ice crystals that formed on different surfaces: "In crevasses, on the roof of the stables, on windows, and so on, countless varied forms are to be seen, each single form corresponding to a particular temperature, humidity, change of temperature, and change of humidity," he wrote. Finding that some crystals measured more than two inches across, Wright tied their size and shape to conditions during their formation. He ultimately concluded that the internal structure of glaciers reflected their founding conditions as well.[50]

From these studies of snow and ice, Wright reexamined the nature of Antarctic icebergs. The barrier snows compacted into true ice, he had determined, so the tabular bergs formed from it would also consist of ice, not layered snow. Of course, "névé-bergs" could form in regions, such as around Cape Adare, where heavy snowfalls built up and moved out to sea before forming ice. "From a distance, it is quite impossible to distinguish the two types," he wrote with Priestley after returning to Britain, "though the majority of tabular icebergs examined closely were found to consist of true ice, though most of it was stratified ice obviously derived from snow." The two researchers also identified "unconformity icebergs," formed with true ice below topped with layers of névé, and noted that some Antarctic glaciers calved nontabular bergs. "At the snouts of such glaciers," Wright and Priestley wrote, "typical icebergs of the more well-known Arctic types are formed in great numbers each year. From these places, they are carried north and west along the coast to join the main pack, where, however, they are overshadowed and fade into relative insignificance beside the great tabular icebergs."[51]

During the Western Journey, Wright had set stakes across the Ferrar Glacier as markers to gauge its movement. In September 1911, for his final outing before the Polar Journey, Scott took two trusted companions, Birdie Bowers and Edgar Evans, along with meteorologist George Simpson on a sledge trip to see if the stakes had moved. "We saw that there had been movement and roughly measured it as about 30 feet," Scott noted. "This is an extremely important observation, the first made on the movement of the coastal glaciers; it is more than I expected to find, but small enough to show that the idea of comparative stagnation was correct."[52]

In his diary, Scott characterized this two-week trip as a useful warm-up for the Polar Journey, but his harshest critic, historian Roland Huntford, later condemned it. "Scott went 150 miles in a senseless direction," Huntford wrote. "He would have done better to go on the Barrier and move fresh seal meat along the road to the south." Certainly the trip was an ordeal: it included man-hauling a sledge in −40°F and a forced march of thirty-five miles in one twenty-four-hour period. "It is not quite clear why they are going," Debenham noted at the time. The most plausible reason was research. Two days before departing, after reviewing the expedition's scientific work and implicitly recognizing the challenge posed by Amundsen, Scott wrote in his diary, "It is a really satisfactory state of affairs all around. If the Southern journey comes off, nothing, not even priority at the Pole, can prevent the Expedition ranking as one of the most important that ever entered the Polar regions." Science would make it so, and thus science must be served. The trip also confirmed Scott's faith in Bowers and Evans for the Polar Journey. "I do not think that harder men or better sledge travellers ever took the trail," he commented.[53]

While Taylor and his party returned west in the spring, Wright headed south on the Polar Journey. The trek began on November 1, 1911, with Scott, Wright, Wilson, Bowers, Edgar Evans, and five others leading pony-sledges across the Ice Barrier. Two dogsleds followed. Motorized tractors had started south six days earlier but broke down so quickly that the so-called Motor Party, consisting of expedition second officer Teddy Evans, stoker William Lashly, engineer Bernard Day, and ship's steward Frederick Hooper, man-hauled its load virtually from the start and increasingly fell behind amid complaints about Evans's leadership. Unlike Shackleton and Amundsen, who each led a single party toward the pole, Scott relied on a large group composed of small parties that fell back in stages until only one was left to man-haul for the pole. Each party carried stores for a series of depots strategically located along the route to resupply returning parties, with the ponies destined to become part of the larder. After a blizzard that kept everyone in camp for five days at the base of Beardmore Glacier and made the surface hellish for hoofed animals, the surviving ponies were shot on December 9 and their carcasses added to the depot. The dogs headed north with their handlers two days later. Scott had already sent back Day and Hooper, leaving three four-person teams to haul heavy sledges up Beardmore Glacier and onto the Polar Plateau.

Wright made it only to the glacier's summit before Scott ordered his party to depot its surplus and return to Cape Evans. "Scott a fool. Teddy goes on," Wright complained in his diary, not realizing that Scott picked Teddy Evans over him to continue because he did not trust Evans to return to base in command of operations. "I must have shown my disappointment," Wright later added, since Scott, "most kindly, softened the blow by pointing out that I would have the responsibility as navigator of the party, of seeing that we did not get lost on the way back. It did soften the blow to a great extent. I was not entirely happy but soon recovered and indeed, probably took this responsibility too seriously."[54] After struggling through the crevasses and pressure ridges midway down Beardmore Glacier, Wright's party limped back to base on reduced rations.

Wright hoped to study ice conditions on the Ice Barrier and Beardmore Glacier during the Polar Journey but had no time for research, not even on the way back. "Much could have been done if a single day's rations could have been available for such scientific work," he observed. "Conditions did not, however, permit this, and the memory fresh in our minds of the five days' dally due to blizzard weather at the foot of the Beardmore Glacier offered us no encouragement to delay."[55] All the polar parties focused on one goal: getting a team to the pole and back. Wright had to settle for collecting some geological specimens and making field notes on glaciers, rock strata, and ice crystals that he saw along the way.

The two remaining parties sledged across the Polar Plateau together for two weeks, coming within 150 miles of the pole at an altitude of more than ten thousand feet. On January 4, Scott sent

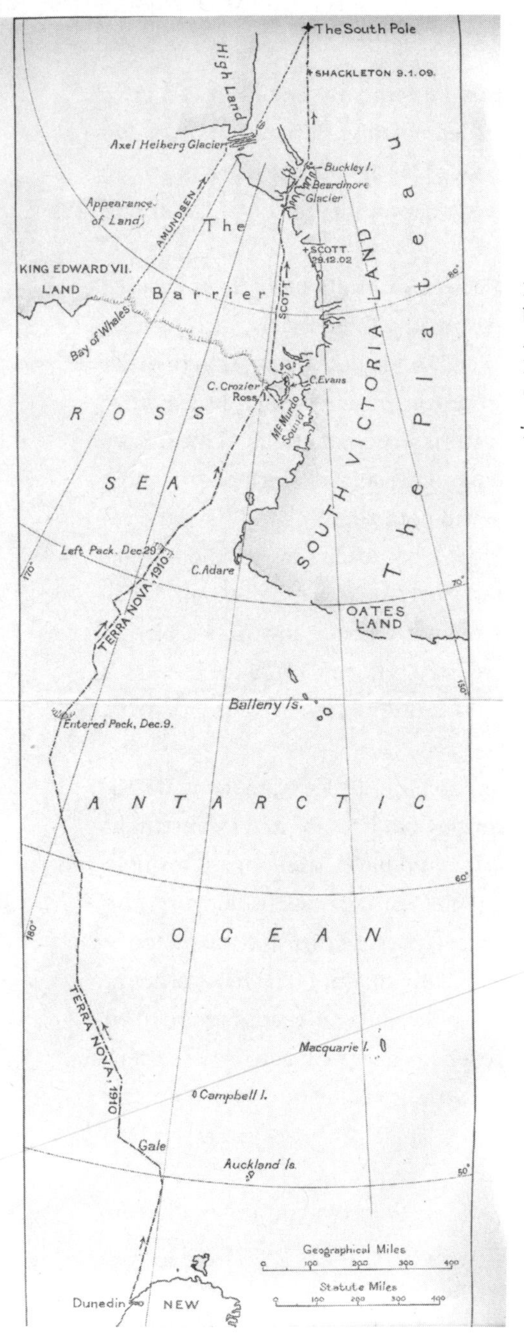

Routes of the parties led by Robert Scott and Roald Amundsen to the South Pole (1911–12), from Apsley Cherry-Garrard's *Worst Journey in the World* (New York, 1922).

Teddy Evans back with Lashly and Thomas Crean, both of whom had been with Scott on the *Discovery* expedition and had returned south with him in hopes of reaching the pole. "They are disappointed," Scott acknowledged. "Poor old Crean wept and even Lashly was affected."[56] But Evans and Lashly had hauled their loads since the tractors broke down, and they were beginning to falter. Crean, Scott allegedly remarked, had "a nasty cough."[57] Their small party barely made it back alive after becoming lost amid the crevasses and pressure ridges on Beardmore and running low on food between depots. Rations calculated at sea level on Wilson's Winter Journey did not suffice at altitude. Prostrate with scurvy, Evans was hauled back the final seventy miles.

Although Evans gave a positive report on the Polar Party's prospects upon his return in late February, the worsening sledging surface that his team encountered should have raised concern. Further, rather than recognize deficits in the food-depot system, leaders at the main base dismissed Evans's illness as an isolated event. Finally, because Amundsen had taken a different route, no one in the British expedition realized that by the time Evans's party turned back in early January, the Norwegians had already reached the pole and returned nearly halfway to their base in near perfect sledging conditions. When *Terra Nova* reached New Zealand on April 2 with the still-prostrate Evans and his report on Scott's progress, it had been nearly a month since Amundsen's *Fram* had docked in Tasmania. Combining the two reports, the world knew that the Norwegians had won the race.

Scott realized Amundsen's triumph as he approached the pole. He now traveled with only four others: Wilson, Bowers, Edgar Evans, and Lawrence "Titus" Oates, an elite cavalry officer and Boer War veteran charged with tending ponies who had also proven himself in sledging. Scott, Wilson, and Bowers were holding up well, though the falling temperature made for hard sledging on icy, granular snow of the type that Wright associated with deep cold. "The surface was bad," Wilson complained, "heavy sandy drifts."[58] Oates's feet were beginning to freeze; Evans had cut his hand badly working on the sledge; Bowers stumped through the snow on foot. Originally part of a support party that had depoted its skis along the way, Bowers walked while the others skied—a difference that could tire even the hardiest sledger.

On January 16, fewer than twenty miles from the pole, they sighted debris left by the Norwegians. "This told the whole story," Scott wrote. "It is a terrible disappointment, and I am very sorry for my loyal companions." Those companions, however, took comfort in what they had done. "It is sad that we have been forestalled by the Norwegians," Bowers wrote, "but I am glad that we have done it by good British man-hauling. This is the greatest journey done by man unaided." For Wilson, solace came partly in science. While the Norwegians won the "race," he noted, "We have done what we came for all the same and as our programme was made out."[59]

"The POLE. Yes, but under very different circumstances from those expected," Scott wrote on January 17. "Great God this is an awful place." His party found the small tent left by Amundsen with a note to Scott and letter for him to take back for the Norwegian king as proof that both groups had reached the goal. Bowers took several rounds of sights to confirm the location, as the Norwegians had done. None of them wanted questions raised about whether they had attained the South Pole such as clouded Robert Peary's claim to the North Pole. The British departed a day later. "We have turned our back now on the goal of our ambition and must face our 800 miles of solid dragging," Scott exclaimed on January 18, "and good-bye to most of the daydreams!"[60]

The surface posed the most relentless problems. It presented the slow, heavy dragging at low, late-summer and early-spring temperatures that Wright would have predicted from his studies of Antarctic snow and ice. "The sun comes out on sandy drifts all on the move in the wind and temp. –20° and gives us an absolutely awful surface with no glide at all for ski or sledge," Wilson wrote on the first day of the return trip. "This in reality is caused by a constant fall of minute snow crystals—*very* minute—sometimes instead of crystal plates the fall is of minute agglomerate spicules like tiny sea urchins."[61] Day after day it was much the same, with the sledgers' diaries filled with complaints about the surface: all pull, no glide, with the sledge runners sometimes sinking so deep into the sandy or crystalline surface that the crossbars plowed through the snow.

They hoped for relief after they left the Polar Plateau on February 7, but the inevitable pitfalls of Beardmore Glacier were followed by even lower temperatures on the Ice Barrier and a persistence of granular snow. "It has been like pulling over desert sand," Scott wrote. "One's heart sinks as the sledge stops dead at some sastrugi behind which the surface sand lies thickly heaped." The temperature now repeatedly plunged below –40°F, which was much lower than Scott expected. Simpson, the expedition's meteorologist, later used available weather data from the region to calculate just how bad it was for the Polar Party. "There was an unusual absence of wind on the south of the Barrier after the middle of February," he concluded. "This allowed radiation to take place unchecked, with the consequence that the temperature became unusually low. This fall of temperature caused a precipitation of water-vapour in the form of ice crystals which appeared in the air and all over the surface, making sledging extremely difficult." Using modern weather data, Antarctic meteorologist Susan Solomon has computed that, during late February and March, the Polar Party faced daily minimum temperatures between 10°F and 20°F below the seasonal norm, which greatly reduced the glide of skis and sledge-runners. "No one in the world would have expected the temperatures and surfaces which we encountered at this time of year," Scott wrote from the Ice Barrier.[62]

Hunger stalked the Polar Party too, even though it traveled on full rations. "Could eat twice what we have, especially at lunch and breakfast," Wilson wrote in a common lament. Until near the journey's end, they reached most depots with food to spare, but the prescribed amount for each meal did not supply enough calories for pulling at altitude over poor surfaces at low temperatures. Midway back, Scott complained, "Thank the Lord we have good food at each meal, but we get hungrier in spite of it." By the time the party reached the Ice Barrier, food had become the major topic of conversation. Then the fuel oil began running low owing to leakage from the stores left in the barrier depots, which meant cold pemmican rather than warm hoosh. "The result is telling on all, but mainly on Oates, whose feet are in a wretched condition," Scott wrote on March 5. No fuel, of course, meant no water.[63]

The unrelenting heavy sledging, low temperatures, and inadequate diet inevitably weakened everyone. For Evans, the serious troubles began when he gashed his hand on December 31. The wound became infected, and his fingers suffered severe and prolonged frostbite. "Evans has dislodged two finger-nails to-night," Scott wrote in late January. "His hands are really bad and to my surprise he shows signs of losing heart over it which makes me much disappointed in him." By the time that the Polar Party reached Beardmore Glacier in early February, Wilson, the party's physician, noted, "Evans' fingers suppurating, nose very bad and rotten looking." By then, Oates's feet had also reached a critical stage. "Titus' toes are blackening," Wilson reported, "and his nose and cheeks are dead yellow."[64] Further, while still on the plateau, Wilson strained a leg tendon so severely that he had to march out of harness for several days, and Scott badly bruised his shoulder in a fall. Although never diagnosed, everyone showed signs of scurvy. Nevertheless, they took time out for geology as they descended the glacier.

Deep snow and ice covered most of the route from the pole to

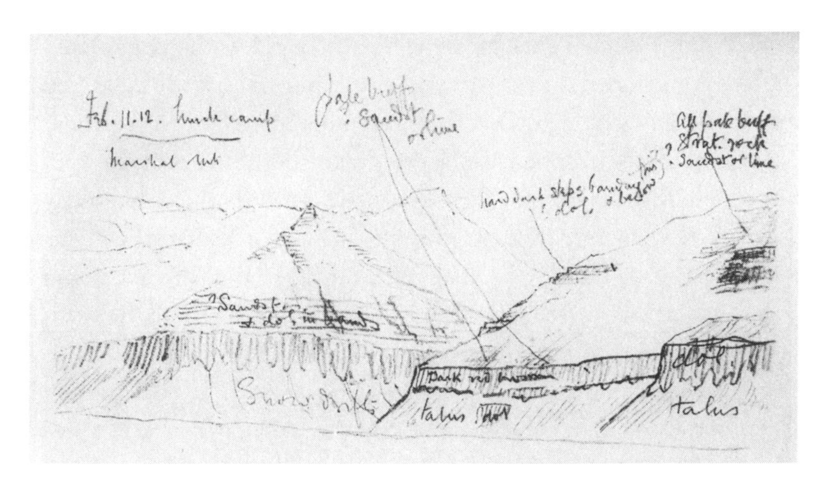

One of Edward Wilson's last sketches, made when he suffered from snow blindness and the Polar Party was running low on supplies, showing the geological strata of exposed rock formations on the north side of Beardmore Glacier, from *Scott's Last Expedition* (New York, 1913).

Ross Island, with the mountains above Beardmore providing the only relief from the featureless monotony. Upon reaching the glacier, the nearly exhausted sledgers steered toward the moraine beneath Mount Buckley. "The moraine was obviously so interesting that . . . I decided to camp and spend the rest of the day geologising," Scott wrote after lunch on February 8. "We found ourselves under perpendicular cliffs of Beacon sandstone, weathering rapidly and carrying veritable coal seams. From the last Wilson, with his sharp eyes, has picked several plant impressions." They took away thirty-five pounds of fossils and rock samples. Those specimens could help salvage meaning from a failed attempt to reach the pole first. "I was very late turning in, examining the moraine after supper," Wilson noted. He sketched the surrounding strata. This break gave the party its last respite.[65]

Evans was the first to go. After floundering down the glacier for a week and falling in several crevasses, he became increasingly disoriented. "He's lost his guts and behaves like an old woman or worse," Oates complained on February 12 in a characteristically gendered remark. Wilson thought that Evans had injured his brain in a fall and by February 16 described him as "sick and giddy and unable to walk even by the sledge on ski."[66] He dropped far behind the others a day later, and when they went back for him, he was on his knees with his gloves off and a wild look in his eyes. By the time they carried him to the tent, he had slipped into a coma and soon died. Leaving the body, they preceded down the glacier and onto the Ice Barrier.

Oates died next. His feet kept getting worse. The cold and the lack of fuel for hot meals contributed to the problem. By the barrier's midpoint, he could no longer pull in harness and soon began holding the party back, which irked the soldier in him. "He asked Wilson if he had a chance this morning," Scott noted on March 10. "In point of fact he has none." A day later, Scott ordered Wilson to supply everyone with a fatal dose of opium, but Oates struggled on with the others for five more days. "He slept through the night before last, hoping not to wake; but he woke in the morning," Scott wrote on March 17. "It was blowing a blizzard. He said, 'I am just going outside and may be some time.' He went out into the blizzard and we have not seen him since." Scott called it "the act of a brave man and an English gentleman."[67] March 17 was Oates's thirty-second birthday. Back at Cape Evans, Tryggve Gran noted this fact in his diary, but Scott did not. Perhaps it was too sad. Perhaps Oates never told him or Scott was beyond caring.

The others marched their last on March 19. Two days earlier, they had left behind everything except the barest essentials and, at Wilson's request, diaries, field notes, and geological specimens. These they carried until the end. Their final camp stood only eleven miles shy of the massive One Ton Depot, stocked with special treats, where a relief party had waited until March 10. Scott's right foot had given out and their fuel was nearly gone, but they retained some hope of getting through. They knew their approximate location. Then a blizzard struck. It lasted at least eight days—or so Scott's diary reported. By then they had eaten all of the remaining food. Some historians have suggested that the blizzard must have ended earlier, but either Wilson and Bowers would not leave Scott or Scott held them back. In any event, they died together, with Wilson and Bowers in an attitude of sleep and Scott between them, his sleeping bag half open and an arm flung across Wilson.

In their last days, the dying men wrote letters to family, friends, and, in Scott's case, the public. Christians of a highly individualistic and intensely personal stripe, Bowers and Wilson wrote of their faith in God and hope for an afterlife. The agnostic, Scott found his faith in the British state wavering so much that his last written words were laced with beseeching pleas for the government and his friends to care for the family he had left behind. Yet his hopes for science, empire, and his own legacy never faded. To the public, he wrote, "I do not regret this journey, which has shown that Englishmen can endure hardships, help one another, and meet death with as great a fortitude as ever in the past." In his letter to his wife, he wrote of their two-year-old son, Peter Markham Scott, "Make the boy interested in natural history if you can; it is better than games." The boy became a renowned ornithologist and environmentalist.[68]

With Scott lost in the south, Teddy Evans invalided to New Zealand, and Campbell clinging to life in the east, leadership of the expedition passed to Edward Atkinson. After *Terra Nova* sailed north without news from the Polar Party, Atkinson kept the expedition going as best as possible. Taylor and Simpson sailed with the ship, leaving Debenham and Wright to continue a scaled-back research program in geology, meteorology, and ice physics. At the first opportunity in spring, Atkinson and Wright led a group south onto the Ice Barrier to search for the remains of the Polar Party. With eyes trained to see anything unusual in the ice, Wright spotted it first. "I saw a small object projecting above the surface on the starboard bow," he later recalled. "It was the 6 inches or so tip of a tent."[69]

Atkinson then took command. "We recovered all their gear," he wrote, "and dug out the sledge with their belongings on it. Amongst these were 35 lbs. of very important geological specimens which had been collected on the moraines of the Beardmore Glacier; at Doctor Wilson's request they had stuck to these up to the very end, even when disaster stared them in the face and they knew that the specimens were so much weight added to what they had to pull. When everything had been gathered up, we covered them with the outer tent and read the burial service. From this time until well into the next day we started to build a mighty cairn above them." The party sledged another twenty miles south in search of Oates's body but found only his sleeping bag.[70]

Apart from the bodies and diaries, the retrieved rocks generated the most comment, then and ever since. Some critics savaged Scott for taking time to collect and expending effort to haul them. Wilson was most eager to do so, however, and his diary expressed his thrill with the fossilized leaf impressions the rocks contained. "Most of the bigger leaves were like beech leaves in shape and venation," he noted, knowing that this tied them to a global flora. Debenham, when he saw them, added, "The 35 lbs. of specimens brought back by the Polar Party from Mt. Buckley contain impressions of fossil plants of late Paleozoic age, some of which a cursory inspection identifies as occurring in other parts of the world." On closer study, the fossils were found to include imprints of the long-sought *Glossopteris* plant, whose presence in Antarctica supported the hypothesis that the southern continents once formed an immense supercontinent. "Meager as it is, the material collected by the Polar Party calls up a picture of an Antarctic land on which it is reasonable to believe were evolved the elements of a new flora that spread in diverging lines over a Paleozoic continent," Cambridge botanist A. C. Seward wrote in 1914. "The discovery of

Glossopteris on the Buckley Island moraine supplies what is needed to bring hypothesis within the range of established fact."[71]

After *Terra Nova* returned to collect the survivors in January 1912, the expedition erected a memorial cross above Hut Point. "It is on the top of Observation Hill," Debenham wrote. "To the west one sees the Western Mountains and Mts Lister and Discovery, while to the south lies White Island and then the blank waste of the Barrier, a dead white plain hardly differing from the sky in colour." Its upright post bore the same line from Tennyson's *Ulysses* that Nansen had applied to Amundsen: "to strive, to seek, to find, and not to yield."[72]

Years later, in a poem about the Antarctic, Debenham wrote, "Men are not old here / Only the rocks are old, and the sheathing ice." The poem went on to comment on the progress of science in that so-called Quiet Land, "The corner is turned: we can see over the brow. / We have sought and found, and it is the land that has yielded."[73] Perhaps so, yet the land took its toll, and the ice continued its relentless push toward the sea. In the 1960s, the ice encasing Framheim, Amundsen's winter quarters, detached from the Ice Barrier, drifted north, and dropped any vestiges of the Norwegians' camp onto the ocean floor. If the movement ascribed to the Ice Barrier by the *Nimrod* and *Terra Nova* expeditions stays constant, the Polar Party's remains, buried beneath a century of accumulated snows, will likewise soon reach the ice front, float free in a tabular burg, and gradually dissolve into the deep southern seas. Given the regional warming that geologists on these expeditions detected, it may have already happened.

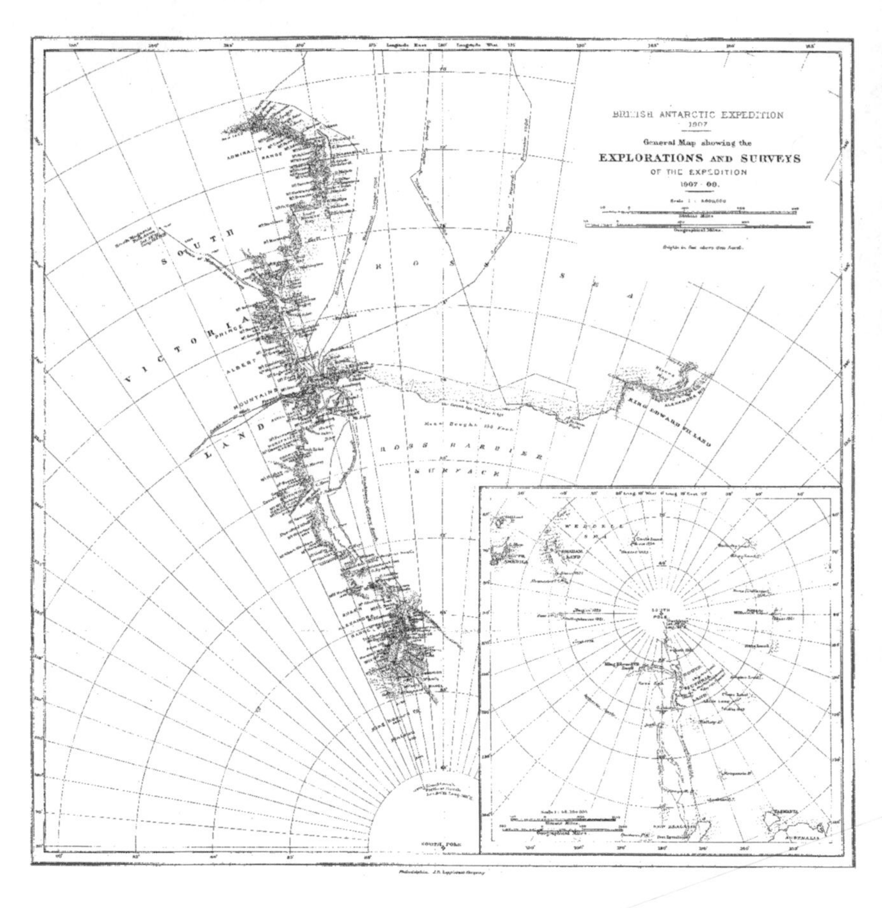

Antarctic map showing what was known in 1910, following the *Discovery* and *Nimrod* expeditions, including the sailing and sledging routes of these expeditions, from Ernest Shackleton's *Heart of the Antarctic* (Philadelphia, 1909).

EPILOGUE

Heroes' Requiem

WHEN THEY LEARNED THAT AMUNDSEN HAD beaten Scott to the South Pole, British commentators typically took solace in science. Their national expedition, they stressed, was more than a dash to the pole; its triumph lay in the findings that Scott would bring back with him. Scott presumably foresaw this, which may explain his decision to geologize on the return trip once he lost priority at the pole. The British saw Scott's effort as more modern and substantive than Amundsen's. After word came that all members of the Polar Party had died on their way back still dragging their field notes and geological specimens, the scientific purpose of the *Terra Nova* expedition initially grew in significance for many in Britain.

Science gave meaning to the death of Scott and his men in a manner that a failed dash to the pole could not. "We have sustained a great and irreparable loss," Clements Markham said on hearing the news of the Polar Party's fate. "No truer and braver men ever served their country, and sacrificed their lives in the cause of science." In his obituary for Scott in the RGS's *Journal,* Markham added: "Robert Scott died as he lived, a faithful and zealous servant of his King and country, a martyr in the cause of science." King George's first message went to the RGS: "I heartily sympathize with the Royal Geographical Society in the loss to science and discovery through the death of these gallant explorers." The Queen Mother, Edward's widow, in a separate message, added, "It may be some slight consolation to know that their purpose has been achieved, and Her Majesty is certain that their splendid and successful undertaking in the cause of science and discovery will for ever be gratefully remembered by the people of this country." Lord Curzon, speaking for the RGS as its president, stated, "I fully anticipate that when Scott's diaries and records come home, and the book is published, it will be found that this expedition is unique in the records of Polar Explorations for its scientific completeness and its results."[1]

News of the deaths dominated British newspapers for weeks and cast a pall over the nation comparable to that occasioned by *Titanic*'s sinking ten months earlier. Countless early news stories pointed to the thirty-five pounds of fossils carried by the Polar Party as evidence of the expedition's commitment to science. "No more pathetic and tragic story has ever unfolded than that of the gallant band of Antarctic explorers whose unavailing heroism now fills the public mind with mingled grief and admiration," the *Times* of London wrote in its lead editorial one day after the news broke. "Nothing in the painful yet inspiring narrative is more touching than the fidelity with which Captain Scott and his comrades, fighting for their very lives with the remorseless forces of Nature, clung with ever increasing peril and weakness to the scientific records and geological specimens which it was the primary object of their expedition to secure. It is thus that they snatched victory out of the jaws of death."[2]

Amundsen, having previously adopted a similar public posture toward Scott's expedition, saw no need to change it. In his already published book about his own polar dash, he had depicted the *Terra Nova* expedition as "designed entirely for scientific research" while his own sought only the pole. The RGS's Hugh Robert Mill had stressed this difference in a review of Amundsen's book published during the same month that word of Scott's death reached Britain. "He claims a verdict of non-interference," Mill wrote of Amundsen, "on the grounds that his motive was altogether different from that of Captain Scott."[3] Upon hearing about Scott's death, Amundsen simply muttered, "Horrible, horrible," and expressed surprise "that such a disaster could overtake a well-organized expedition." These sentiments echoed those of Ernest Shackleton, Robert Peary, Fridtjof Nansen, and other polar explorers. The Swedish adventurer Sven Hedin eulogized Scott: "He has reached his goal. He has served his science." After listing the expedition's research efforts, the British science journal *Nature* concluded, "The nation has reason to be proud of these men, who have laid down their lives in the pursuit of geographical knowledge."[4]

The lamentations soon turned from science, largely in response to Scott's dying "Message to the Public," which gave a different meaning to the tragedy. It said nothing about research results but offered a poignant picture of fortitude in extreme conditions. "Had we lived," Scott wrote, "I should have had a tale to tell of the hardihood, endurance, and courage of my companions which would have stirred the heart of every Englishman." Their deaths told it better. According to the *Spectator,* a conservative weekly, Scott's message was "as calm, noble, and courageous as any words ever written by a man in the face of death." Markham called it "one of the finest pieces of writing in our language." At noon on February 14, 1913, four days after news of his death reached London, Scott's message was read to one and a half million British schoolchildren to coincide with a memorial service for the Polar Party in Saint Paul's Cathedral attended by the king and prime minister. Titus Oates's final words as recorded by Scott—"I may be some time"—rang with Scott's final message as lessons in courage and service. "Self-sacrifice is the keynote," the *Contemporary Review* concluded. By choosing death to save his colleagues, even though they, too, later died, Oates "had set up before the eyes of a faltering generation a standard of life that cannot be measured by death. He was, indeed a very gallant gentleman, and so was each man of his company."[5]

With this, almost without anyone's noticing, the story had subtly changed. "No one was interested in the expedition's scientific achievements," historian Stephanie Barczewski wrote based on her analysis of the response. "Instead, it was the story of the tragic death of Scott and his four companions that dominated the press."[6] Even the RGS moderated its message. "I will not pause to consider what the scientific results of this expedition will be," Lord Curzon told the membership at a memorial meeting in late February. "It does not matter very much that [Antarctica] was once united to the Australasian and South American continents; that it once enjoyed a more temperate climate; or that forests flourished there which have left traces of coal beneath the ice and the snow. But it does matter to us and to the entire world a great deal that men have been found in this, as in earlier and perhaps more virile ages, to run great risks for a great idea, to count life itself as dust in the balance compared with supreme human endeavour, and to meet death without repining even on the threshold of victory and fame."[7] Returning to Britain in command of the expedition and serving as its spokesman, Teddy Evans declared, "By the disaster to our Expedition we have lost five of the best men the country has produced, but we have set a lesson that is far more valuable than any scientific results, the lesson to young English men and women to do their duty."[8]

In 1913, with war looming and amid fears of imperial decline, Britain more urgently needed inspiring examples of self-sacrificing service than of Antarctic research. The Admiralty promptly declared that Scott and Edgar Evans would be deemed "killed in action," with military pensions provided to family members. "Their story will long be remembered with honor by the Navy," the *Times* reported.[9] The army paid similar tribute to Oates. Following Teddy Evans's public address about the expedition at Queen's Hall in June 1913, Winston Churchill, in the chair as first lord of the Admiralty, declared that the Polar Party's example gave "to us, in these islands, and to our kith and kin in the Colonies and Dominions" confidence "that if ever other tests should be applied to the naval system upon which the British Empire depends, the officers and men of the sea service, and their comrades in the Army, will not fail this country or the Empire in the hour of need."[10]

To bolster morale during World War I, the British military showed films from the expedition to troops in the field. Kathleen Scott reported receiving letters from soldiers acknowledging the inspiration provided by her husband's sacrifice. Edward Atkinson, the expedition surgeon who led the party that found Scott's body in 1913, went to war with the troops. "Many a man," he wrote of his wartime experiences, "has been wounded and laid out in No Man's Land since then to close his life in pain and discomfort. Perhaps the example set by the fate of the Polar Party and the way in which they met it may have served its purpose as an example to the rest."[11]

Many scientists from the *Discovery, Nimrod,* and *Terra Nova* expeditions served with distinction in the war. Both Charles Wright and Raymond Priestley received the Military Cross for gallantry in action. Enlisting at age fifty-eight, Edgeworth David engineered the mining of enemy positions in the Battle of Messines and rose to the rank of lieutenant-colonel. Louis Bernacchi received awards from both the British and American military for his naval service. After leading his own expedition to the Antarctic coast west of Cape Adare from 1911 to 1914, Douglas Mawson served as a major assigned to the Ministry of Munitions. Hartley Ferrar and George Simpson held technical positions with British forces in the Middle East. Frank Debenham, a major, nearly died from wounds suffered while on patrol in Greece. Wright, Priestley, Mawson, David, Simpson, and Debenham went on to successful careers in science or scientific administration.

Most of the navy sailors and officers on the expeditions also went to war, and several died in action. Two officers, Charles Royds and Teddy Evans, became admirals; Victor Campbell attained the rank of captain and further fame as a hero at sea by using his ship to ram and sink a surfaced German U-boat. Shackleton, preparing for another Antarctic expedition when the war began, was directed by the Admiralty to proceed: his ship was crushed in the Weddell Sea ice before reaching land. The eighteen-month homeward journey on pack ice and in an open boat across the storm-tossed Southern Ocean secured Shackleton's place in the pantheon of British explorers. In 1922, he died of a heart attack on the outbound leg of his final expedition south. Six years later, Amundsen perished in the Arctic when his plane crashed on a rescue mission. Markham had died in 1916, victim of his traditional ways. Ignoring the concerns of his servants, he persisted in reading by candlelight in his draped canopy bed long after his house was wired with electricity. Bedridden with gout, he died when the candle ignited the bedclothes. "The Coroner," the *Times* of London noted, "said Sir Clements Markham, being an old-fashioned man, preserved old-fashioned things."[12]

Two world wars kept the heroic interpretation of the Polar Party's ordeal relevant for a generation, but it lost currency in their aftermath. Stripped of science and equated with man-hauling to the pole, Scott's *Terra Nova* expedition came to stand for little more than relentless perseverance in the face of inevitable defeat. Its means appeared Victorian and its goal arbitrary, especially after Amundsen efficiently, effectively, and seemingly effortlessly won the race by using dogs. Over time, the ideals that drove Scott south on foot became blurred with those that drove the Great War's gallant carnage. In the hindsight of two awful wars and the dismemberment of Britain's global empire, such ideals seemed quixotic at best and dangerous at worst, particularly when they cost the lives of others. Modern efficiency and effective leadership came into vogue—values commonly associated with Amundsen and Shackleton. In Britain and the United States, Scott was eclipsed by Shackleton, whose public image benefitted from his flexible leadership style and success in bring his men back alive from virtually hopeless situations. Beginning with a popular 1979 book about the race to the Pole by Roland Huntford, *Scott and Amundsen,* which was hypercritical of Scott's choices, Scott became widely seen as a bungler. "Thanks to Huntford," Barczewski wrote in 2007, "no other single word has come to be more associated with Scott."[13]

The British historian Eric Hobsbawm famously declared that "the long nineteenth century" did not end until August 1914, with the opening of World War I. He could have chosen March 1912, when Robert Falcon Scott died. Historical demarcations gain meaning only as later generations reflect on earlier events. At the time, one day, year, or century simply leads to another. Even the guns of August 1914 at first struck most observers as merely the outbreak of another of the wars of empire that had pockmarked the nineteenth century. Only reflection and the enormity of the ensuing conflict made them symbolic. Contemporary British commentators often depicted Victoria's death in 1901 as the end of a once glorious but slowing fading century and Edward's ascension as the start of a revitalized, forward-looking one. Scott's *Discovery* expedition, whose sendoff was one of Edward's early public acts, appeared to symbolize a new century in which science would link arms with discovery in the peaceful expansion of empire. The guns of August leveled these pretentions and, in later histories, transformed Edward's decade into a hollow coda to his mother's century. The view of Scott's polar exploits evolved accordingly.

Unlike World War I, Scott's death did not cause direct change. In opening the first chapter of his 1987 history of the era, Hobsbawm dismissed Scott as "hapless" and declared that his attempt to reach the pole neither "had nor was intended to have the slightest practical consequences."[14] In saying so, he expressly ignored the research activities of Scott's *Terra Nova* expedition, which both had and were intended to have ongoing scientific significance. Yet his comments reflect the general sense that *Terra Nova,* like Amundsen's adventure, was all about the pole. As the view of Scott's efforts as hapless, hidebound, and hollow gained currency, his death and those of his comrades acquired a meaning much like that of the guns of August: fading echoes of the Victorian age as it gave way to modern times. By 2000, after larger portraits of Scott were moved to storage, his only image on exhibit in London's National Portrait Gallery sat in a display case devoted to Victorian explorers and empire builders. Shackleton's portrait, in contrast, hung in the early-twentieth-century section.

But when science is restored to the equation, all three British Antarctic expeditions of the prewar period become modern and forward-looking enterprises. Of course, people still go to Antarctica for adventure and, like Amundsen, to bag poles, but the continent has become fundamentally a place of science. Hundreds of researchers go there every year to pursue lines of investigation pioneered by the *Discovery, Nimrod,* and *Terra Nova* expeditions. Of the two expedition leaders, Scott developed a greater personal commitment to science over the course of his two expeditions even though, in David, Mawson, Priestley, and James Murray, Shackleton attracted an abler team of scientists. All three expeditions conducted significant research that, in fields ranging from climate change and paleontology to marine biology and glaciology, helped to shape the twentieth-century view of Antarctica and its place in the global system of nature. The returning researchers worked to keep the study of Antarctic science alive through two world wars and into the 1950s, when academic attention again turned to Antarctic topics during the International Geophysical Year. Although the focus on heroic but hapless man-hauling turned Scott into a Victorian stereotype, giving due recognition to research, the British expeditions of the Edwardian age should represent precursors of the current era in Antarctic science.

NOTES

CHAPTER 1. "THREE CHEERS FOR THE DOGS"

1. Hugh Robert Mill, *The Siege of the South Pole: The Story of Antarctic Exploration* (London: Alston Rivers, 1905), 71; James Cook, *A Voyage towards the South Pole and round the World,* vol. 1 (London: W. Strahan, 1777), 268.

2. James Clark Ross, *A Voyage of Discovery and Research in the Southern and Antarctic Regions,* vol. 1 (London: Murray, 1847), 117; Mill, *Siege of the South Pole,* 273.

3. Ross, *Voyage in Antarctic Regions,* 218.

4. Robert F. Scott, *The Voyage of the "Discovery,"* vol. 2 (New York: Scribner's, 1905), 79.

5. E. H. Shackleton, *The Heart of the Antarctic: Being the Story of the British Antarctic Expedition, 1907–1909,* vol. 1 (Philadelphia: Lippincott, 1909), 347–48.

6. Mary Shelley, *Frankenstein; or, The Modern Prometheus, The Novels and Selected Works of Mary Shelley,* ed. Nora Crook, vol. 1 (London: Pickering, 1996), 169.

7. Richard Collinson, *Journal of H.M.S. "Enterprise" on the Expedition in Search of Sir John Franklin's Ships by Behring Strait* (London: Clowes, 1889), 368; George H. Richards to Sophia Cracroft, Jan. 14, 1877, Scott Polar Research Institute (SPRI) Archives, MS 248/462/24.

8. Roald Amundsen, *My Life as an Explorer* (Garden City, NY: Doubleday, 1928), 2; Roald Amundsen, *The Northwest Passage: Being the Record of a Voyage of Exploration of the Ship "Gjoa," 1903–1907,* vol. 1 (London: Constable, 1908), 4.

9. Amundsen, *My Life,* 27.

10. Fridtjof Nansen, eulogy for Roald Amundsen, 1928, quoted in Ror Bomann-Larsen, *Roald Amundsen* (London: Sutton, 2006), 32; Amundsen, *My Life,* 33.

11. Peder Ristvedt in Bomann-Larsen, *Amundsen,* 41.

12. Roald Amundsen to Leon Amundsen, Nov. 1, 1906, in ibid., 55; Leon Amundsen to Roald Amundsen, in ibid., 55.

13. Gerald Christy to Roald Amundsen, in ibid., 57; Amundsen, *My Life,* 62.

14. John Scott Keltie to Roald Amundsen, in Bomann-Larsen, *Amundsen,* 57.

15. George T. Goldie, in Roald Amundsen, "To the North Magnetic Pole and through the North-West Passage," *Geographical Journal* 29 (1907), 513.

16. Roald Amundsen in ibid., 485.

17. Fridtjof Nansen in ibid., 514–15.

18. "The Monthly Record," *Geographical Journal* 29 (1907): 567.

19. Plan of the expedition, in "A Proposed North Polar Expedition," *Nature* 79 (1909): 412.

20. Roald Amundsen to Leon Amundsen, in Bomann-Larsen, *Amundsen,* 71.

21. Amundsen, *My Life,* 65.

22. Fridtjof Nansen, Nov. 17, 1911, in Kathleen Scott, *Self-Portrait of an Artist: From the Diaries and Memoirs of Lady Kennet* (London: Murray, 1949), 102.

23. Roald Amundsen to Fridtjof Nansen, Aug. 22, 1910, in Roland Huntford, *Scott and Amundsen* (London: Hodder, 1979), 294.

24. Roald Amundsen to Robert F. Scott, Oct. 3, 1909, in ibid., 317 (a common version of this widely quoted telegram).

25. Roald Amundsen, *The South Pole: An Account of the Norwegian Antarctic Expedition in the "Fram," 1910–1912,* 2 vols. (London: Murray, 1913), 1:42–44.

26. Ibid., 52.

27. Clements Markham to John Scott Keltie, Dec. 12, 1910, Royal Geographical Society Archives, MS CB7 [1881–1910].

28. Amundsen, *South Pole,* 2:121, 132.

29. Ibid., 133.

30. Ibid., 135, 143–44.

31. "The South Pole Discovered," *New York Times,* Mar. 8, 1912, 12.

32. Scott, Mar. 6, 1912, *Self-Portrait,* 107; "The Discovery of the South Pole," *Spectator* 108 (1912): 428.

33. "England Retains Its Faith in Scott," *New York Times,* Mar. 9, 1912, 4; "The South Pole Won," *Times* (London), Mar. 9, 1912, 9; "Captain Amundsen's Achievement," ibid., Mar. 9, 1912, 5; "London's Only Report," *New York Times,* Mar. 8, 1912, 2.

34. Peter Scott, Mar. 11, 1912, in Scott, *Self-Portrait,* 108. A first child, Peter was born ten months before his father left for Antarctica in 1910.

35. "Scott's Last Expedition," *Bookman* 45 (December 1913): 66.

36. Herbert G. Ponting, in "News of Capt. Scott Expected in Spring," *New York Times,* Dec. 22, 1912, C6.

37. "Diary of the Week," *Nation* (London), Apr. 6, 1912, 3; "Capt. Scott in the Antarctic," *Athenaeum,* Apr. 13, 1912, 416; "English Disappointment," *New York Times,* Apr. 1, 1912, 1; "Notes," *Nature* 89 (1912): 112–17.

38. "Geography and the Empire," *Times* (London), May 21, 1912, 6.

39. "British Sneer at Amundsen," *New York Times,* Sept. 10, 1912, 4.

40. Roald Amundsen to Don Pedro Christophersen, in Bomann-Larsen, *Amundsen,* 121; Clements Markham, Nov. 7, 1912, in Scott, *Self-Portrait,* 115; Scott, Nov. 15, 1912, in ibid., 115.

41. Lord Curzon, "The Norwegian South Polar Expedition," *Geographical Journal* 41 (1913): 15, 16.

42. Amundsen, *My Life,* 72.

43. Ibid., 72; Roald Amundsen to Leon Amundsen, December 1912, in Bomann-Larsen, *Amundsen,* 136.

44. "British Association," *Times* (London), Sept. 10, 1912, 47; Frank Mundell, *Stories of South Pole Adventure* (London: Sunday School Union, 1912), 179.

CHAPTER 2. A COMPASS POINTING SOUTH

1. This and the following quotations are from "[Instructions] by the Commissioners," in James Clark Ross, *A Voyage of Discovery and Research in the Southern and Antarctic Regions, during the Years 1839–43,* vol. 1 (London: Murray, 1847), xxii, xxiv, xxvi–xxvii.

2. William Whewell, "Report on Electricity, Magnetism, and Heat," in British Association for the Advancement of Science, *Report of the Fifth Meeting* (London: Murray, 1836), 1.

3. "Report of the Sub-Committees," in British Association for the Advancement of Science, *Report of the First Meeting* (London: Murray, 1833), 52; James D. Forbes, "Report on Meteorology," in British Association for the Advancement of Science, *Report of the Second Meeting* (London: Murray, 1833), 258; "Magnetism," in British Association for the Advancement of Science, *Report of the Fifth Meeting* (London: Murray, 1836), xxi.

4. "Reports Requested," in British Association for the Advancement of Science, *Report of the Eighth Meeting* (London: Murray, 1838), xxi; Roderick Impey Murchison, "President's Address," in British Association for the Advancement of Science, *Report of the Eighth Meeting* (London: Murray, 1839), xxxv.

5. John Herschel, "Report on Resolutions," in British Association for the Advancement of Science, *Report of the Ninth Meeting* (London: Murray, 1840), 36, 38–39.

6. Ibid., 37–39. In his 1845 presidential address to the BAAS, after the triumphal return of Ross's expedition, Herschel would revisit the issue of state funding for magnetic research: "It is the pride and boast of an Englishman to pay his taxes cheerfully when he feels assured of their application to great and worthy objects." John Herschel, "President's Address," in British Association for the Advancement of Science, *Report of the Fifteenth Meeting* (London: Murray, 1848), xxxi.

7. J. F. W. Herschel, "Report of the Joint Physical and Meteorological Committee," in Ross, *Voyage,* viii, xv–xvi.

8. John Herschel, "Diary," Oct. 15, 1838, Royal Society (RS) Archives, MS 583; Instructions, in Ross, *Voyage,* xxvi.

9. W. Vernon Harcourt, "President's Address," in British Association for the Advancement of Science, *Report of the Ninth Meeting* (London: Murray, 1839), 1–2, 4; William Whewell, *History of the Inductive Sciences,* 3rd ed., vol. 2 (New York: Appleton, 1882), 231 (quotation added in the 2nd edition of 1845). Herschel made a virtually identical comment in Herschel, "President's Address," xxx.

10. Ross, *Voyage,* 176, 182–83.

11. J. Ross, "Extracts of a Dispatch," *Times* (London), Jan. 7, 1842, 3.

12. Ibid.; Ross, *Voyage,* 189.

13. Ross, "Extracts," 3; Ross, *Voyage,* 217.

14. Ross, "Extracts," 3.

15. Ibid.

16. "Captain Ross and the Antarctic Expedition," *Times* (London), Sept. 22, 1842, 6; "The Antarctic Expedition," ibid., Sept. 11, 1843, 6.

17. Edward Sabine, quoted in F. J. Evans, "The Magnetism of the Earth," *Proceedings of the Royal Geographical Society of London* 22 (1878): 205. Sabine's tables, analysis and maps of magnetic data from the Antarctic regions collected from 1840 to 1845 are in Edward Sabine, "Contributions to Terrestrial Magnetism: No. XI," *Philosophical Transactions of Royal Society of London* 158 (1868): 371–416.

18. Evans, "Magnetism," 216.

19. Clements R. Markham, "The Present Standpoint of Geography," *Geographical Journal* 2 (1893): 483.

20. Erasmus Ommanney, "Antarctic Discovery," *Proceedings of the British Association* (55th annual meeting), n.s., 7 (1885): 759.

21. John Murray, in "The British Association," *Times* (London), Sept. 16, 1885, 6.

22. Robert F. Scott, *The Voyage of the "Discovery,"* 2 vols. (New York: Scribner's, 1905), 1:27.

23. Clements Markham, *Antarctic Obsession: A Personal Narrative of the Origins of the British National Antarctic Expedition, 1901–1904* (Alburgh, UK: Bluntisham Books, 1986), 2.

24. Clements Markham, *The Lands of Science: A History of Arctic and Antarctic Exploration* (Cambridge: Cambridge University Press, 1921), 447; Markham, *Antarctic Obsession*, 4; Francis Spufford, *I May Be Some Time: Ice and the English Imagination* (New York: Picador, 1997), 274. For a range of views on Markam's sexual preferences from two leading Scott biographers, compare Roland Huntford, *Scott and Amundsen* (London: Hodder, 1979), 125 ("Though married, with a daughter, Markham was a homosexual. He sometimes went south to indulge his proclivities safe from criminal prosecution"), with David Crane, *Scott of the Antarctic: A Life of Courage and Tragedy* (New York: Knopf, 2006), 62 ("There seems little doubt that there was a homosexual strain in [Markham's] passionate attachment to youth. . . . There is not a shred of evidence, however, . . . that his predilection for youth was anything more than that").

25. Markham, *Lands of Silence*, 445.

26. Clements R. Markham, "The Need for an Antarctic Expedition," *Nineteenth Century* (October 1895): 710; "The Antarctic Expedition," *Times* (London), Mar. 5, 1895, 10 (Markham's speech to the Imperial Institute). For similar comments, see Clements R. Markham, "The Antarctic Expedition," Apr. 10, 1895, Royal Geographical Society (RGS) Archives, MS AA1/2/9 (private printing of Markham's speech to the Royal United Service Institute).

27. Markham, *Antarctic Obsession*, 62.

28. John Murray, "The Renewal of Antarctic Exploration," *Geographical Journal* 3 (1894): 25.

29. Ibid., 14, 23. The printed address included a long appendix by Neumayer explaining the critical need for magnetic readings from the Antarctic regions.

30. Markham, *Lands of Silence*, 444.

31. Clements R. Markham, "Opening Address," *Report of the Sixth International Geographical Congress* (London: Murray, 1896), 17; A. W. Greely, "Borchgrevink and Antarctic Exploration," *Century Magazine*, January 1896, 431; "Discussion," *Report of the Sixth Congress*, 176.

32. Greely, "Borchgrevink," 431; C. E. Borchgrevink, "The Voyage of the 'Antarctic' to Victoria Land," *Report of the Sixth Congress*, 174.

33. Llewellyn W. Longstaff to Clements Markham, Mar. 28, 1899, RGS Archives, MS AA 2/1/9.

34. "The Antarctic Ship *Discovery*," *Times* (London), Mar. 20, 1901, 13.

35. Clements R. Markham, "An Antarctic Expedition," June 12, 1898, RGS Archives, MS AA2/1/2, 3; Clements R. Markham, "The Antarctic Expeditions," *Geographical Journal* 14 (1899): 481.

36. Clements R. Markham to William Speirs Bruce, May 16, 1900, Scott Polar Research Institute (SPRI) Archives, MS 441/16.

37. "The National Antarctic Expedition: Instructions to the Commander," *Geographical Journal* 18 (1901): 154; Markham, *Lands of Silence*, 453; "Instructions," 154.

38. "Instructions," 156.

39. Robert F. Scott to Hannah Scott, Feb. 24, 1903, SPRI Archives, MS 1542/8/1; L. C. Bernacchi, *Saga of the "Discovery"* (London: Blackie, 1938), 44; Robert F. Scott to Archibald Geikie, Oct. 19, 1904, RS Archives, MS 257/1318.

40. Scott, *Voyage of the "Discovery,"* 2:131; Clements Markham, "The Antarctic Expedition," *Verhandlungen des siebenten Internationalen Geographen-Kongresses* (Berlin: Kuhl, 1901), 625.

41. Reginald Koettlitz, quoted by Albert Armitage in Scott, *Voyage of the "Discovery,"* 2:133; Albert B. Armitage, *Two Years in the Antarctic: Being a Narrative of the British National Antarctic Expedition* (London: Arnold, 1905), 169.

42. Armitage, *Two Years in the Antarctic,* 173, 182; Markham, *Lands of Silence,* 465; Robert F. Scott, "Brief Summary of Proceedings," Feb. 23, 1903, RS Archives, MS 591/1 (5); William Speirs Bruce, "Farthest South," *London Magazine,* June 1909, 356.

43. Hugh Robert Mill, *An Autobiography* (London: Longmans, 1951), 149.

44. "Instruction for Northern Sledge-Party," in E. H. Shackleton, *The Heart of the Antarctic: Being the Story of the British Antarctic Expedition, 1907–09,* vol. 2 (Philadelphia: Lippincott, 1909), 73.

45. Douglas Mawson, in *Mawson's Antarctic Diaries,* ed. Fred Jacka and Eleanor Jacka (London: Unwin-Hyman, 1988), 15–16, 25.

46. Ibid., 23–24, 27, 36–37.

47. "Professor David's Narrative," in Shackleton, *Heart of the Antarctic,* 127; Mawson, *Antarctic Diaries,* 34–35.

48. "David's Narrative," 177–78.

49. Douglas Mawson, "Magnetic Observations," in Shackleton, *Heart of the Antarctic,* 384; "David's Narrative," 181.

50. Mawson, *Antarctic Diaries,* 45.

51. L. C. Bernacchi, "Farthest South," *Travel and Exploration* 1 (1909): 337; "Royal Congratulations," *London Magazine,* June 1909, 350; Hugh Robert Mill, "Lieutenant Shackleton's Achievement," *Geographical Journal* 33 (May 1909): 572.

CHAPTER 3. THE EMPIRE'S MAPMAKER

1. "The National Antarctic Expedition: Instructions to the Commander," *Geographical Journal* 18 (1901): 154–55.

2. Robert A. Stafford, *Scientist of Empire: Sir Roderick Murchison, Scientific Exploration and Victorian Imperialism* (Cambridge: Cambridge University Press, 1989), 211.

3. Roderick Impey Murchison, "Address to the Royal Geographical Society of London," *Journal of the Royal Geographical Society of London* 22 (1852): cxxiii–cxxiv.

4. David Livingstone, in Ian Cameron, *To the Farthest Ends of the Earth: 150 Years of World Exploration by the Royal Geographical Society* (New York: Dutton, 1980), 87–88.

5. David Livingstone, *Missionary Travels and Researches in South Africa* (London: Ward, [1900?]), 3.

6. David Livingstone, "Extracts of Letters from the Rev. David Livingstone," *Journal of the Royal Geographical Society of London* 20 (1850): 138–41.

7. Ibid.

8. Roderick Impey Murchison, quoted by Thomas Maclear in George Seaver, ed., *David Livingstone: His Life and Letters* (New York: Harper, 1957), 244.

9. Livingstone, *Missionary Travels,* 445.

10. Thomas Maclear in Seaver, *David Livingstone,* 280; James MacQueen, "Notes on the Geography of Central Africa, from the Researches of Livingstone," *Journal of the Royal Geographical Society of London* 26 (1856): 109.

11. Roderick Impey Murchison to David Livingstone, May 23, 1856, and Oct. 2, 1855, in Seaver, *Livingstone,* 265.

12. Livingstone, *Missionary Travels,* 449.

13. John Murray to David Livingstone, Jan. 5, 1856, in Seaver, *Livingstone,* 265. For *Origin of Species,* Darwin paid half of the cost and received half of the profits.

14. Charles Dickens, "Thanks to Doctor Livingstone," *Household Words* 17 (1858): 121.

15. "Dr. Livingstone's African Discoveries," *Times* (London), Dec. 16, 1956, 10.

16. Ibid.; Cameron, *To the Farthest Ends of the Earth,* 102.

17. "The East African Mission," *Times* (London), Jan. 20, 1963, 5.

18. "Livingstone," *New York Herald,* July 2, 1872, 1.

19. H. M. Stanley, "Central Africa and the Congo Basin: Or, the Importance of the Scientific Study of Geography," *Journal of the Manchester Geographical Society* 1 (1885): 13, 24; Hugh Robert Mill, *The Record of the Royal Geographical Society, 1830–1930* (London: Royal Geographical Society, 1930), 129.

20. William Gladstone, quoted by Roderick Impey Murchison, in discussion of T. G. Montgomerie, "Report of the Trans-Himalayan Explorations Made during 1868," *Proceedings of the Royal Geographical Society of London* 14 (1860–70): 214; Robert A. Stafford, "Scientific Exploration and Empire," in *The Oxford History of the British Empire: Nineteenth Century,* ed. Andrew Porter (Oxford: Oxford University Press, 1999), 300.

21. Clements R. Markham, "View of the Progress of Geographical Discovery," *Encyclopaedia Britannica,* 9th ed., vol. 10 (New York: Werner, 1898), 195.

22. Clements R. Markham, *The Fifty Years' Work of the Royal Geographical Society* (London: Murray, 1881), 90; Albert H. Markham, *The Life of Sir Clements R. Markham* (London: Murray, 1917), 228; "The Arctic Expedition," *Times* (London), Mar. 15, 1874, 4.

23. Markham, *Fifty Years' Work,* 92; Markham, *Life of Markham,* 245.

24. Markham, "Geographical Discovery."

25. E. A. Reeves, ed., *Hints to Travellers: Scientific and General,* 9th ed., 2 vols. (London: Royal Geographical Society, 1906), 1:1; John Coles, ed., *Hints to Travellers: Scientific and General,* 8th ed., vol. 1 (London: Royal Geographical Society, 1901), 1.

26. Douglas W. Freshfield and W. J. L. Wharton, eds., *Hints to Travellers: Scientific and General,* 7th ed. (London: Royal Geographical Society, 1893), 1.

27. Reeves, *Hints,* 2:290.

28. Robert F. Scott, *The Voyage of the "Discovery,"* 2 vols. (New York: Scribner's, 1905), 2:239–40 (quoting diary).

29. Robert F. Scott, "Lecture," 1, undated manuscript, Scott Polar Research Institute (SPRI) Archives, MS 336/16/16; similar text in Robert F. Scott, "The National Antarctic Expedition," *Times* (London), Nov. 8, 1904, 6.

30. Scott, *Voyage of the "Discovery,"* 2:264 (quoting diary); Albert B. Armitage, *Two Years in the Antarctic: Being a Narrative of the British National Antarctic Expedition* (London: Arnold, 1905), 182.

31. James Clark Ross, *A Voyage of Discovery and Research in the Southern and Antarctic Regions,* vol. 1 (London: Murray, 1847), 221; Scott, *Voyage of the "Discovery,"* 1:167–68.

32. Scott, *Voyage of the "Discovery,"* 1:170.

33. Ibid., 1:171, 173 (quoting diary), 191.

34. Ibid., 1:198–99, 201–2; see also Armitage, *Two Years,* 59.

35. Clements Markham, *The Lands of Science: A History of Arctic and Antarctic Exploration* (Cambridge: Cambridge University Press, 1921), 453; Clements Markham, *Antarctic Obsession: A Personal Narrative of the Origins of the British National Antarctic Expedition, 1901–1904* (Alburgh, UK: Bluntisham Books, 1986), 47, 55.

36. Edward Wilson, *Diary of the "Discovery" Expedition to the Antarctic Regions, 1901–1904* (New York: Humanities Press, 1967), 150–51.

37. Scott, *Voyage of the "Discovery,"* 2:20, 23 (quoting diary).

38. Wilson, *Diary of the "Discovery,"* 226, 227–28.

39. Scott, *Voyage of the "Discovery,"* 2:69; Wilson, *Diary of the "Discovery,"* 216.

40. Scott, *Voyage of the "Discovery,"* 2:52 (quoting diary).

41. Wilson, *Diary of the "Discovery,"* 230; Scott, *Voyage of the "Discovery,"* 2:53 (quoting diary).

42. Scott, *Voyage of the "Discovery,"* 2:31–32, 79–80 (quoting diary).

43. Ibid., 2:125; "Antarctic Exploration," *Times* (London), Sept. 10, 1904, 7; Clements R. Markham, "First Year's Work of the National Antarctic Expedition," *Geographical Journal* 22 (1903): 18.

44. Robert Scott to Hannah Scott, Feb. 24, 1903, SPRI Archives, MS 1542/8/1-7; Scott, *Voyage of the "Discovery,"* 2:121 (quoting diary). See also Robert Scott to Hannah Scott ("Our own journey to the South was, of course, the severest, the distance travelled was the longest, the time longest, and our food allowance the least. For Wilson and myself, saddled as we were with an invalid for three weeks at the end, it was especially trying").

45. "London, Saturday, September 10, 1904," *Times* (London), Sept. 10, 1904, 9; Letter to Member of RGS Council, quoted in Reginald Pound, *Scott of the Antarctic* (New York: Coward-McCann, 1966), 92; Markham, *Lands of Silence,* 476; Hannah Scott to Robert Scott, May 11, 1903, SPRI Archives, MS 1542/8/1-7.

46. E. H. Shackleton, "A New British Antarctic Expedition," *Geographical Journal* 29 (1907): 331.

47. E. H. Shackleton, *The Heart of the Antarctic: Being the Story of the British Antarctic Expedition, 1907–1909,* 2 vols. (Philadelphia: Lippincott, 1909), 1:297 (quoting diary).

48. Ibid., 1:299–300 (quoting diary).

49. Ibid., 1:310–11, 315, 333 (quoting diary).

50. Ibid., 1:339–40 (quoting from diary Dec. 28–29), 344 (quoting diary).

51. E. H. Shackleton, "Some Results of the British Antarctic Expedition, 1907–9," *Geographical Journal* 34 (1909): 490.

52. Shackleton, *Heart of the Antarctic,* 1:348 (quoting diary); Emily Shackleton to H. R. Mill, Aug. 16, 1922, SPRI Archives, MS 100/104/1-66 (black-edged stationery marking her husband's death on a later expedition).

53. F. W. Everett, "Shackleton," *Pearson's Weekly,* Apr. 8, 1909, 818.

54. "Progress of the World," *Review of Reviews* 40 (1909): 8.

55. Shackleton, *Heart of the Antarctic,* 2:229.

56. H. Bartle Frere, "Address to the Royal Geographical Society," *Proceedings of the Royal Geographical Society* 18 (1874): 511.

CHAPTER 4. IN *CHALLENGER'S* WAKE

1. C. Wyville Thomson, *The Depths of the Sea* (New York: Macmillan, 1873), 1–2.

2. Ibid., 2.

3. Ibid., 410, 411.

4. Ibid., 495, 471.

5. C. Wyville Thomson, *The Atlantic,* 2 vols. (New York: Harper, 1878), 1:21.

6. Thomson, *Depths of Sea,* 401.

7. "Report of the Committee," Nov. 30, 1871, in Thomson, *Atlantic,* 1:75.

8. "Her Majesty's Ship Challenger," *Times* (London), Dec. 5, 1872, 3.

9. "The Scientific Orders of the *Challenger,*" *Nature* 7 (1873): 191.

10. Ibid.

11. N. H. Moseley, *Notes by a Naturalist: His Account of Observations Made during the Voyage of H.M.S. "Challenger"* (London: Murray, 1892), 1.

12. Ibid., 501–2; "Dinner to the *Challenger* Staff," *Nature* 14 (1876): 239.

13. Wyville Thomson, "The *Challenger* Expedition," *Nature* 7 (1873): 386.

14. "The Ocean Survey Expedition," *Times* (London), Dec. 7, 1872, 12.

15. Jack Skylight, "A Letter without Much Rhyme and with a Little Reason," *Nature* 9 (1874): 304.

16. "The Livingstone and Challenger Expeditions," *Times* (London), Dec. 10, 1872, 7.

17. Joseph Hooker, "Present Condition of the Royal Society," *Nature* 11 (1874): 177.

18. C. Wyville Thomson, "The *Challenger* Expedition," *Nature* 14 (1876): 492; "The *Challenger,*" *Times* (London), May 26, 1876, 8.

19. C. Wyville Thomson, "The *Challenger* Expedition," *Nature* 11 (1874): 117; T. H. Huxley, "Notes from the *Challenger,*" *Nature* 12 (1875): 316.

20. "The Challenger," *Times* (London), Nov. 30, 1877, 3; "The First Fruits of the *Challenger* Expedition," ibid., Dec. 13, 1880, 9.

21. John Murray, in George Murray, ed., *The Antarctic Manual for the Use of the Expedition of 1901* (London: Royal Geographical Society, 1901), 242, 245.

22. Thomson, "*Challenger* Expedition," 495.

23. Thompson, *Atlantic,* 2:302.

24. Wyville Thomson, "Notes from the *Challenger,*" *Nature* 14 (1876): 15; Thomson, "*Challenger* Expedition," 495.

25. Moseley, *Notes by a Naturalist,* 202; C. Wyville Thomson, "On the Conditions in the Antarctic," *Nature* 15 (1876): 120.

26. George Nares, in H. Bartle Frere, "Address to the Royal Geographical Society," *Proceedings of the Royal Geographical Society of London* 18 (1974): 546; "Her Majesty's Ship *Challenger,*" *Times* (London), May 26, 1974, 7.

27. Thomson, "On the Conditions," 120–21.

28. Ibid., 120.

29. W. B. Carpenter, "Ocean-Circulation," *Contemporary Review* 26 (1875): 587; William B. Carpenter, "Ocean Currents," *Nature* 9 (1874): 424.

30. Hugh Robert Mill, *The Siege of the South Pole: The Story of Antarctic Exploration* (London: Alston Rivers, 1905), 350.

31. John Murray, "The Revival of Antarctic Exploration," *Geographical Journal* 3 (1894): 18–19.

32. Ibid., 16–17.

33. Ibid., 25; Clements R. Markham to Albert Markham, Aug. 11, 1899, National Maritime Museum (NMM) Archives, MS MRK/46.

34. Clements R. Markham, *Antarctic Obsession: A Personal Narrative of the Origins of the British National Antarctic Expedition, 1901–1904* (Alburgh, UK: Bluntisham Books, 1986), 128, 136; Clements R. Markham, "Notes on Proposed alterations by the Joint Committee,"

n.d. [February or March 1901], Royal Geographical Society (RGS) Archives, MS AA 1/6/6; Clements R. Markham to Albert Markham, Nov. 10, 1899, NMM Archives, MS MRK/46 (120).

35. Edward B. Poulton to Fellows of the Royal Society, repr. in *Nature* 64 (1901): 83; Markham, *Antarctic Obsession*, 133.

36. Markham, "Notes"; Clements R. Markham, "Antarctic Expedition: Memorandum on Dr. Gregory" [Feb. 14, 1901], RGS Archives, MS AA 4/1/17.

37. Edward B. Poulton to Albert B. Armitage, Jan. 27, 1901, repr. in Albert B. Armitage, *Cadet to Commodore* (London: Cassell, 1925), 135; Armitage, *Cadet to Commodore*, 139.

38. Archibald Geikie, quoted in Poulton to Fellows, 85; J. W. Gregory to Arthur Rücker, Jan. 9, 1901, repr. in ibid., 84.

39. Markham, *Antarctic Obsession*, 100, 102; George Murray to Hugh Robert Mill, June 26, 1899, Scott Polar Research Institute (SPRI) Archives, MS 100/83/1-11.

40. "Instructions to Commander," *Geographical Journal* 18 (1901): 154–55.

41. Ibid.

42. Robert F. Scott to Hugh Robert Mill, Dec. 17, 1901, SPRI Archives, 100/100/2; T. V. Hodgson, "On Collecting in Antarctic Seas," in National Antarctic Expedition (NAE), *Natural History*, 6 vols. (London: British Museum, 1907), 2:10.

43. Robert F. Scott, *The Voyage of the "Discovery,"* 2 vols. (New York: Scribner's, 1905), 1:129; T. H. Baughman, *Pilgrims on the Ice: Robert Falcon Scott's First Antarctic Expedition* (Lincoln: University of Nebraska Press, 1999), 50.

44. Hodgson, "On Collecting," 10; Erich von Drygalski, "The Oceanographical Problems of the Antarctic," in *Problems of Polar Research*, ed. W. L. G. Joerg (New York: American Geographical Society, 1928), 277.

45. T. V. Hodgson, "Crustacea," in NAE, *Natural History*, 5:1.

46. Quotations in the two following paragraphs are from Hodgson, "On Collecting," 5, 9–10.

47. Edward T. Browne, "Medusa," in NAE, *Natural History*, 5:2.

48. Scott, *Voyage of the "Discovery,"* 2:434.

49. Ibid., 1:275, 2:431.

50. Ibid., 1:276.

51. E. H. Shackleton, "Observations," *South Polar Times* 1 (April 1902): 21.

52. Markham, *Antarctic Obsession*, 74.

53. W. N. Shaw, "Preface," in National Antarctic Expedition (NAE), *Meteorology*, 2 vols. (London: Royal Society, 1908–13), 1:vi; Scott, *Voyage of the "Discovery,"* 1:320.

54. R. H. Curtis, "Notes on the Observations of Barometric Pressure," in NAE, *Meteorology*, 1:489.

55. L. C. Bernacchi, *Saga of the "Discovery"* (London: Blackie, 1938), 87 (referring to the sledge journey with Royds, Bernacchi writes, "We, of course, saw no new land, nothing but the barren, level surface of the barrier, with few undulations to break the monotony").

56. Curtis, "Notes," 478.

57. Archibald Geikie, "Preface," in NAE, *Meteorology*, 2:4; Shaw, "Preface," xii.

58. Beau Riffenburgh, *Shackleton's Forgotten Expedition: The Voyage of the "Nimrod"* (New York: Bloomsbury, 2004), 140.

59. James Murray, "On Collecting at Cape Royds," in British Antarctic Expedition (BAE), *Biology*, vol. 1 (London: Heinemann, 1910), 10, 13.

60. Ibid., 9, 13; James Murray, "Editorial Note," to Charles Hedley, "Mollusca," in BAE, *Biology*, 2.

61. Scott, *Voyage of the "Discovery,"* 2:429; R. H. Curtis, "Discussion of the Observations of the Direction and Force of the Winds," in NAE, *Meteorology,* 1:495.

62. E. H. Shackleton, *The Heart of the Antarctic: Being the Story of the British Antarctic Expedition, 1907–1909,* 2 vols. (Philadelphia: Lippincott, 1909), 1:170–71.

63. Ibid., 1:177; T. W. Edgeworth David, "The Ascent of Mount Erebus," in *Aurora Australis,* ed. E. H. Shackleton (Cape Royds: Privately printed, 1908), [13].

64. Shackleton, *Heart of the Antarctic,* 1:187 (quoting report by David and Adams); David, "Ascent of Erebus," [27].

65. Eric S. Marshall, "Diary of British Antarctic Expedition," Mar. 10–11, 1908, SPRI Archives, MS 1456/8, 7.

66. Ernest Shackleton, "Erebus," BAE, in *The Antarctic Book: Winter Quarters, 1907–1909* (London: Heinemann, 1909), 21.

67. Shackleton, *Heart of the Antarctic,* 1:200 (quoting report by David and Adams).

68. T. W. Edgeworth David and Raymond E. Priestley, *British Antarctic Expedition 1907–9 Reports on the Scientific Investigations: Geology,* vol. 1 (London: Heinemann, 1914), 14, 25, 29.

69. T. W. Edgeworth David and J. B. Adams, "Meteorology; A Summary of Results," in Shackleton, *Heart of the Antarctic,* 2:406–7.

70. Shackleton, *Heart of the Antarctic,* 1:199 (quoting report by David and Adams).

CHAPTER 5. TAKING THE MEASURE OF MEN

1. G. S. Nares, *Narrative of a Voyage to the Polar Sea,* vol. 1 (London: Sampson Low, 1878), 374 (from Markham's diary).

2. Ibid., 1:375 (from Markham's diary).

3. Sherard Osborn, "On the Exploration of the North Polar Region," *Proceedings of the Royal Geographical Society* 9 (1865): 62 (statement of Clements R. Markham).

4. Ibid., 43–44, 48.

5. Ibid., 52–54; Clements Markham, "Address to the Royal Geographical Society," *Geographical Journal* 14 (1899): 12 (quoting Osborn's speech in slightly different language).

6. Osborn, "On the Exploration," 63 (statement of Dr. Donnet), 57–58 (quoting Sabine letter); Lady Franklin to Roderick Murchison, Apr. 6, 1865, *Proceedings of the Royal Geographical Society* 9 (1865): 148 (read to society following Osborn's address).

7. Benjamin Disraeli to Henry Rawlinson, Nov. 17, 1874, in *Proceedings of the Royal Geographical Society* 19 (1874): 39; "Report of the Arctic Committee of the Admiralty," ibid., 19 (1875): 350.

8. "London, Tuesday, October 31, 1876," *Times* (London), Oct. 31, 1876, 7; "Royal Geographical Society," ibid., Nov. 11, 1874, 7 (quoting Wilson); "WAITING TO BE WON," *Punch,* June 5, 1875, 240; C. R. Markham, "On the Best Route for North Polar Exploration," *Proceedings of the Royal Geographical Society* 9 (1865): 143; "The Arctic Expedition," *Times* (London), May 23, 1875, 8.

9. "The Arctic Expedition," *Times* (London), May 31, 1875, 10; "The Arctic Expedition," ibid., May 21, 1875, 10 (mayor's statement, "Arctic Expedition," May 31, 1875, 10 (queen's message and following).

10. Nares, *Narrative of Voyage,* 139; A. H. Markham, "On Sledge Travelling," *Proceedings of the Royal Geographical Society* 21 (1876): 113.

11. Markham, "Sledge Travelling," 113–14.

12. Ibid., 114–15; Nares, *Narrative of Voyage,* 1:395 (from Markham's diary); Markham, "Sledge Travelling," 119.

13. Nares, *Narrative of Voyage,* 358, 373, 377 (from Markham's diary).

14. Ibid., 386.

15. "The Arctic Expedition," *Times* (London), Oct. 28, 1876, 8 (reprint of telegram).

16. "London, Saturday, October 28, 1876," *Times* (London), Oct. 28, 1876, 9; David McGonigal and Lynn Woodworth, *Antarctica and the Arctic: The Complete Encyclopedia* (London: Firefly, 2001), 510 (quotations from newspaper); Geo. Henry Richards to Editor, *Times* (London), Nov. 13, 1876, 8.

17. George Henry Richards, "Introduction," in Nares, *Narrative of Voyage,* xl.

18. "Naval Mismanagement," *Times* (London), Dec. 4, 1876, 12 (reprint of Saturday Review editorial); "London, Friday, December 1, 1876," ibid., Dec. 1, 1876, 9; Nares, *Narrative of Voyage,* 395 (statement of Markham).

19. Sylvanus Urban, "Table Talk," *Gentleman's Magazine* 239 (1876): 767; "Possibilities," *Times* (London), Nov. 6, 1876, 12 (reprint of *Spectator* editorial); "Consolations," ibid. (reprint of *Saturday Review* editorial); "The Arctic Expedition," ibid., Dec. 9, 1876, 10 (reprint of Nares statement).

20. Henry Rawlinson, "New Polar Expedition," *Proceedings of the Royal Geographical Society* 19 (1874): 40; Albert H. Markham, *The Life of Sir Clements R. Markham* (London: Murray, 1917), 245 (quoting Clements Markham on the successful results of the Arctic Expedition).

21. Francis Galton, *Hereditary Genius: An Inquiry into Its Laws and Consequences* (New York: Appleton, 1870), 362.

22. Clements Markham, *Antarctic Obsession: A Personal Narrative of the Origins of the British National Antarctic Expedition, 1901–1904* (Alburgh, UK: Bluntisham Books, 1986), 4–5.

23. Ibid., 13; "The National Antarctic Expedition," *Times* (London), May 29, 1900, 3.

24. Markham, *Antarctic Obsession,* 13–15, 2.

25. Mandell Creighton to his niece, Dec. 16, 1899, in Louise Creighton, *Life and Letters of Mandell Creighton,* vol. 2 (London: Longmans, 1904), 405; L. S. Amery, ed., *The* Times *History of the War in South Africa, 1899–1900,* vol. 1 (London: Sampson Low, 1900), 11.

26. George C. Brodrick, "A Nation of Amateurs," *Nineteenth Century* 48 (1900): 525; Sidney Low, "The Breakdown of Voluntary Enlistment," ibid., 47 (1901), 366; "Lord Rosebery on Questions of Empire," *Times* (London), Nov. 17, 1800, 19.

27. Karl Pearson, *National Life from the Standpoint of Science* (London: Black, 1905), 11 (reprint of 1900 lecture); "Reconstruction or Catastrophe?" *National Review* 36 (1900): 330; Andrew Carnegie, "British Pessimism," *Nineteenth Century* 49 (1901): 901; Harold E. Gorst, "The Blunder of Modern Education," ibid., 847; "London, Saturday, September 10, 1904," *Times* (London), Sept. 10, 1904, 9.

28. J. A. Hussey to H. R. Mill, July 27, 1922, Scott Polar Research Institute (SPRI) Archives, MS 100/49/1-5 (claims to quote Shackleton).

29. H. T. Farrar, "Diary," Feb. 19, 1902, SPRI Archives, MS 1153/1; Edward Wilson, *Diary of the "Discovery" Expedition to the Antarctic Regions, 1901–1904* (New York: Humanities Press, 1967), 130; Robert F. Scott, *The Voyage of the "Discovery,"* 2 vols. (New York: Scribner's, 1905), 1:226.

30. Scott, *Voyage of the "Discovery,"* 1:226.

31. Ibid., 229.

32. David Crane, *Scott of the Antarctic: A Life of Courage and Tragedy* (New York: Knopf, 2006), 147; Scott, *Voyage of the "Discovery,"* 1:237–39.

33. Scott, *Voyage of the "Discovery,"* 1:238–39; Robert F. Scott, "Diary," Mar. 12, 1902, SPRI Archives, MS 352/1/2.

34. Scott, *Voyage of the "Discovery,"* 1:239 (quoting Barne's report), 241–43, 252.

35. Ibid., 1:261–62 (from Royds's report).

36. Reginald Skelton, "Diary," opposite entry for Oct. 18, 1902, SPRI Archives, MS 342/1/2.

37. Scott, *Voyage of the "Discovery,"* 1:270–72; James Duncan, "Diary," Apr. 1, 1902, SPRI Archives, MS 1415; Wilson, *Diary of the "Discovery,"* 173.

38. Scott, *Voyage of the "Discovery,"* 1:360; Wilson, *Diary of the "Discovery,"* 162; Scott, *Voyage of the "Discovery,"* 1:273.

39. Scott, *Voyage of the "Discovery,"* 1:379–80, 472.

40. Ibid., 2:327.

41. Ibid., 2:399.

42. "Return of the National Antarctic Expedition," *Geographical Journal* 24 (1904): 379–81.

43. Clements Markham, "The Antarctic Expedition," *Geographical Journal* 23 (1904): 552; Clements R. Markham, "Address to the Royal Geographical Society," ibid., 22 (1903): 13; "The Antarctic Expedition," *Times* (London), Apr. 2, 1904, 4.

44. Bruton Galleries, *"Discovery" Antarctic Expedition* (London: Bruton Galleries, 1904), 15–18, 29–30, 57–68 (quotations on 30, 64). Copy in SPRI Archives, MS 366/16/1–5.

45. "The National Antarctic Expedition," *Times* (London), Nov. 8, 1904, 6.

46. *Daily Graphic* reporter, quoted in Crane, *Scott,* 279; Markham, "Antarctic Expedition," 549–50; "London," *Times* (London), Sept. 10, 1904, 9; "Antarctic Exploration," ibid., 7; Elspeth Huxley, *Scott of the Antarctic* (London: Weidenfeld, 1977), 174 (quoting Admiralty official).

47. "Return of the National Antarctic Expedition," 381 (Markham).

48. E. H. Shackleton, *The Heart of the Antarctic: Being the Story of the British Antarctic Expedition, 1907–1909,* vol. 1 (Philadelphia: Lippincott, 1909), 1; Roland Huntford, *Shackleton* (New York: Atheneum, 1986), 117; J. Scott Keltie to R. F. Scott, Feb. 18, 1907, Royal Geographical Society (RGS) Archives, RFS/4a.

49. J. Scott Keltie to R. F. Scott, Feb. 18, 1907; J. A. McIlroy, interviewed by James Fisher, Aug. 16, 1955, SPRI Archives, MS SR/I/A/10.

50. "New British Expedition to the South Pole," *Times* (London), Feb. 12, 1907, 12; "By Car to the South Pole," *Car,* Oct. 23, 1907, 398.

51. Shackleton, *Heart of the Antarctic,* 342.

52. E. H. Shackleton to Emily Shackleton, Feb. 12, 1907, SPRI Archives, MS 1537/2/12/15; P. W. Marony, "Antarctic Photography," *Australian Photographic Journal,* Nov. 12, 1903, 325; Clements R. Markham, "Notes of Shackleton's Book," December 1909, 1, RGS Archives, EHS 2 (handwritten manuscript).

53. J. Scott Keltie to E. H. Shackleton, Apr. 1, 1909, RGS Archives, MS EHS 2; "Lieut. Shackleton's Antarctic Expedition," *Geographical Journal* 33 (1909): 485; Beau Riffenburgh, *Shackleton's Forgotten Expedition: The Voyage of the "Nimrod"* (New York: Bloomsbury, 2004), 280–89 (quoting the *Sketch* and the *Sphere*).

54. "Leading Articles in the Reviews," *Review of Reviews* 40 (1909): 240; E. R. Wethey, "Geography Notes Up-to-Date," *Practical Teacher* 29 (1909): 561.

55. R. F. Scott to J. Scott Keltie, Feb. 20, 1907, Mar. 28, 1908, RGS Archives, MS RFS/4a; R. F. Scott, quoted in "The British Antarctic Expedition," *Geographical Journal* 34 (1909): 126; "Mr. Shackleton and the Savage Club," *Times* (London), Jun. 21, 1909, 6 (quoting Scott);

Tom Crean, in "E. L. Atkinson's Report of the British Antarctic Expedition, 1910–13," 1 (pt. 1, sec. a), SPRI Archives, MS 280/28/1, 1.

56. "A British Antarctic Expedition," *Times* (London), Sept. 13, 1909, 9, 12.

57. "Luncheon to British Antarctic Expedition, 1910," *Geographical Journal* 36 (1910): 22–23.

58. William Lashly, *Under Scott's Command: Lashly's Antarctic Diaries* (London: Gollancz, 1969), 118.

59. Scott, *Voyage of the "Discovery,"* 1:467–68.

60. F. Leopold McClintock, "On Arctic Sledge-Travelling," in *The Antarctic Manual for the Use of the Expedition of 1901,* ed. George Murray (London: Royal Geographical Society, 1901), 293; Clements R. Markham, "Memorandum for the Landing Party Committee," n.d., National Maritime Museum Archives, MRK/46 (106); Clements Markham, "The Antarctic Expeditions," *Verhandlungen des siebenten Internationalen Geographen-Kongresses,* Berlin, 1899 (Berlin: Kuhl, 1901), 625; Pelham Aldrich to R. F. Scott, Sept. 26, 1903, SPRI Archives, MS 366/15.

61. Robert Falcon Scott, *Journals: Captain Scott's Last Expedition,* ed. Max Jones (Oxford: Oxford University Press, 2005), 189; Lashly, *Scott's Command,* 121; Scott, *Journals,* 315; Edward Wilson, *Diary of the "Terra Nova" Expedition to the Antarctic, 1910–1912* (New York: Humanities Press, 1967), 213; Scott, *Journals,* 345; H. R. Bowers to Kathleen Scott, Oct. 27, 1911, SPRI Archives, MS 1488/2 (vol. 1).

62. Scott, *Journals,* 185, 209.

CHAPTER 6. MARCH TO THE PENGUINS

1. "The Plans for Antarctic Exploration," *Nature* 60 (1899): 203 (response of Balfour); "The National Antarctic Expedition," *Geographical Journal* 18 (1901): 27 (statement of Edward VII).

2. Clements R. Markham, "Address to the Royal Geographical Society," *Geographical Journal* 22 (1903): 13.

3. J. W. Gregory, "The Work of the National Antarctic Expedition," *Nature* 63 (1901): 612.

4. George Murray, ed., *The Antarctic Manual for the Use of the Expedition of 1901* (London: Royal Geographical Society, 1901), 288.

5. "The Royal Society's Antarctic Meeting," *Geographical Journal* 11 (1898): 422 (statement of Sclater); "Deputation to the Government," ibid., 14 (1899): 196 (statement of Hooker).

6. P. L. Sclater to Edward Wilson, June 1, 1900, in George Seaver, *Edward Wilson of the Antarctic: Naturalist and Friend* (London: Murray, 1933), 72; Edward Wilson, in ibid., 75; Edward Wilson, *Diary of the "Discovery" Expedition to the Antarctic Regions, 1901–1904* (New York: Humanities Press, 1967), 89 (diary for Dec. 31, 1901).

7. Robert F. Scott, *The Voyage of the "Discovery,"* 2 vols. (New York: Scribner's, 1905), 1:94; Wilson, *Diary of the "Discovery,"* 50–51.

8. Wilson, *Diary of the "Discovery,"* 77–80; Scott, *Voyage of the "Discovery,"* 1:105.

9. Wilson, *Diary of the "Discovery,"* 53, 80–81.

10. Scott, *Voyage of the "Discovery,"* 1:124; Wilson, *Diary of the "Discovery,"* 90–91.

11. Wilson, *Diary of the "Discovery,"* 93–95, 106.

12. Scott, *Voyage of the "Discovery,"* 1:149; Wilson, *Diary of the "Discovery,"* 100.

13. Edward A. Wilson, "Some Notes on Penguins," *South Polar Times* 1 (July 1902): 6;

Edward A. Wilson, "Aves," in National Antarctic Expedition, *Natural History*, vol. 2 (London: British Museum, 1907), 3–4; Wilson, *Diary of the "Discovery,"* 110; Wilson, "Aves," 4.

14. C. Wyville Thomson, *The Depths of the Sea* (New York: Macmillan, 1873), 8–9.

15. Ernst Haeckel, *The Riddle of the Universe at the Close of the Nineteenth Century* (New York: Harpers, 1901), 81.

16. P. L. Sclater, "Notes on the Emperor Penguin," *Ibis* 30 (1888): 325, 328.

17. Wilson, "Some Notes," 3.

18. "Events of the Month," *South Polar Times* 1 (April 1902): 11.

19. Wilson, *Diary of the "Discovery,"* 132; Thomas Kennar, "Told at One Bell," *South Polar Times* 1 (April 1902): 11–12.

20. Robert F. Scott, *Voyage of the "Discovery,"* 2:5.

21. Wilson, *Diary of the "Discovery,"* 206; Wilson, "Aves," 23, 25.

22. Wilson, *Diary of the "Discovery,"* 207.

23. Wilson, "Aves," 5.

24. Scott, *Voyage of the "Discovery,"* 2:8.

25. Wilson, *Diary of the "Discovery,"* 267.

26. Ibid., 288; Wilson, "Aves," 6.

27. Wilson, *Diary of the "Discovery,"* 294.

28. Wilson, "Aves," 31.

29. Ibid., 310, 306.

30. Ibid., 308, 309.

31. Edward A. Wilson to Edward T. Wilson, in George Seaver, *Edward Wilson of the Antarctic: Naturalist and Friend* (London: Murray, 1933), 146.

32. Wilson, "Aves," 11–12; Wilson, *Diary of the "Discovery,"* 294 (apparent diary source for this portion of the "Aves" account); Wilson, "Aves," 11–12.

33. Clements Markham, *Antarctic Obsession: A Personal Narrative of the Origins of the British National Antarctic Expedition, 1901–1904* (Alburgh, UK: Bluntisham Books, 1986), 86.

34. Wilson, "Aves," 11.

35. E. H. Shackleton, "A New British Antarctic Expedition," *Geographical Journal* 29 (1907): 331.

36. James Murray, "On Collecting at Cape Royds," in British Antarctic Expedition, *Reports on the Scientific Investigations: Biology*, vol. 1 (London: Heinemann, 1910), 4.

37. Ibid.; Jules Cardot, "Musci," in ibid., 77.

38. James Murray, "Biology," in E. H. Shackleton, *The Heart of the Antarctic: Being the Story of the British Antarctic Expedition, 1907–1909*, vol. 2 (Philadelphia: Lippincott, 1909), 236; Murray, "On Collecting," 6.

39. Murray, "Biology," 239–40; James Murray and George Marsden, *Antarctic Days* (London: Melrose, 1913), 81 (from chapter by Murray).

40. R. N. Rudmose Brown, "Some Results of the British Antarctic Expedition, 1907–1909," *Internationale Revue der gesamten Hydrobiologie und Hydrographie* 4 (1911): 372.

41. Murray, "On Collecting," 1; E. H. Shackleton, "Some Results of the British Antarctic Expedition, 1907–9," *Geographical Journal* 34 (1909): 484.

42. Wilson, "Aves," 31.

43. Robert Falcon Scott, *Journals: Captain Scott's Last Expedition*, ed. Max Jones (Oxford: Oxford University Press, 2005), 236.

44. Cossar Ewart, "Report," in Apsley Cherry-Garrard, *The Worst Journey in the World: Antarctic, 1910–1913*, vol. 1 (New York: Doran, 1922), 301; Griffith Taylor, *With Scott: The Silver Lining* (New York: Dodd, 1916), 271; Edward Wilson, "Lecture on Penguins," in ibid., 244.

45. Scott, *Journals*, 236; Taylor, *With Scott*, 271.

46. Scott, *Journals*, 237, 249 (corrected to Scott's original text).

47. Edward Wilson, *Diary of the "Terra Nova" Expedition to the Antarctic, 1910–1912* (New York: Humanities Press, 1967), 146, 147; Cherry-Garrard, *Worst Journey*, 237–38.

48. Wilson, *Diary of "Terra Nova,"* 149; Cherry-Garrard, *Worst Journey*, 260.

49. Cherry-Garrard, *Worst Journey*, 262.

50. "The Winter Journey," in R. F. Scott, *Scott's Last Expedition*, vol. 2 (New York: Dodd, 1913), 18 (from Cherry-Garrard's diary for July 15, 1911).

51. Ibid., 22 (from Cherry-Garrard's diary for July 19, 1911); Wilson, *Diary of "Terra Nova,"* 153; "Winter Journey," 25.

52. Cherry-Garrard, *Worst Journey*, 268; Wilson, *Diary of "Terra Nova,"* 154; Cherry-Garrard, *Worst Journey*, 272.

53. "Winter Journey," 31 (from Cherry-Garrard's diary for July 22, 1911); Wilson, *Diary of "Terra Nova,"* 156; Cherry-Garrard, *Worst Journey*, 279 (quoting Bowers's diary); Wilson, *Diary of "Terra Nova,"* 157.

54. Wilson, *Diary of "Terra Nova,"* 157; see also "Winter Journey," 35 (from Wilson's Report to Scott).

55. Wilson, *Diary of "Terra Nova,"* 158.

56. Ibid.; Cherry-Garrard, *Worst Journey*, 293; Wilson, *Diary of "Terra Nova,"* 158.

57. Cherry-Garrard, *Worst Journey*, 297–99; "Winter Journey," 52.

58. Taylor, *With Scott*, 285; Scott, *Journals*, 255.

59. "Winter Journey," 48 (from Cherry-Garrard's account).

60. Edward W. Nelson, "Our Bill," *South Polar Times* 3 (September 1911): 98.

61. Scott, *Journals*, 259.

62. R. F. Scott to J. J. Kinsey, Oct. 28, 1911, Scott Polar Research Institute Archives, MS 761/8/13-34; Scott, *Journals*, 466.

63. Charles S. Wright, *Silas: The Antarctic Diaries and Memoir of Charles S. Wright* (Columbus: Ohio State University Press, 1993), 28; Taylor, *With Scott*, 273.

64. Edward Wilson, "The Barrier Silence," *South Polar Times* 3 (October 1911): 151.

CHAPTER 7. DISCOVERING A CONTINENT'S PAST

1. John Murray, "The Renewal of Antarctic Exploration," *Geographical Journal* 3 (1894): 10, 12.

2. Ibid., 24; Duke of Argyll, "The Renewal of Antarctic Exploration—Discussion," *Geographical Journal* 3 (1894): 29; Clements Markham, *The Lands of Science: A History of Arctic and Antarctic Exploration* (Cambridge: Cambridge University Press, 1921), 442.

3. G. W. Gregory, "The Work of the National Antarctic Expedition," *Nature* 63 (1901): 610–11.

4. Ibid., 611; Georg von Neumayer, "Royal Society's Antarctic Meeting," *Geographical Journal* 11 (1898): 421.

5. Charles Darwin to Joseph Hooker, Aug. 6, 1881, in *The Life and Letters of Charles Darwin*, ed. Francis Darwin, vol. 3 (London: Murray, 1887), 248; Gregory, "Work of the Expedition," 612; P. L. Sclater, "Royal Society's Antarctic Meeting," *Geographical Journal* 11 (1898): 422.

6. L. Fletcher, "Instructions for Collecting Rocks and Minerals," in *The Antarctic Manual for the Use of the Expedition of 1901*, ed. George Murray (London: Royal Geographical Society, 1901), 202–3; W. T. Blanford, "Geology," in ibid., 176.

7. Edward B. Poulton, "The National Antarctic Expedition," *Nature* 64 (1901): 83.

8. Robert F. Scott to Hannah Scott, Feb. 24, 1903, Scott Polar Research Institute (SPRI) Archives, MS 1542/8/1; Clements Markham, *Antarctic Obsession: A Personal Narrative of the Origins of the British National Antarctic Expedition, 1901–1904* (Alburgh, UK: Bluntisham Books, 1986), 102.

9. H. T. Ferrar, "Some Remarks on the Geology of the Neighborhood," *South Pole Times* 1 (June 1902): 22; H. T. Ferrar, "Summary of the Geological Observations Made during the Cruise of the S.S. 'Discovery,' 1901–1904," in Robert F. Scott, *The Voyage of the "Discovery,"* vol. 2 (New York: Scribner's, 1905), 442.

10. Ferrar, "Some Remarks," 23; H. T. Ferrar, "Report on the Field-Geology of the Region Explored during the 'Discovery' Antarctic Expedition, 1901–4," in National Antarctic Expedition, *Natural History: Geology,* vol. 1 (London: British Museum, 1907), 87.

11. Ferrar, "Some Remarks," 24.

12. Albert B. Armitage, *Two Years in the Antarctic* (London: Arnold, 1905), 127.

13. Ferrar, "Summary," 452; H. T. Ferrar, "On Igneous Rocks," *South Polar Times* 2 (August 1903): 22; Robert F. Scott to Hannah Scott, Feb. 28, 1903.

14. Scott, *Voyage of the "Discovery,"* 248–49.

15. Ibid., 250.

16. Ibid., 298.

17. Ferrar, "Report," 40–43.

18. Ibid., 43.

19. David Crane, *Scott of the Antarctic: A Life of Courage and Tragedy* (New York: Knopf, 2006), 440.

20. Scott, *Voyage of the "Discovery,"* 259.

21. Ibid., 287–90.

22. Ibid., 291.

23. Ibid., 293.

24. E. H. Shackleton, "A New British Antarctic Expedition," *Geographical Journal* 29 (1907): 331.

25. Raymond E. Priestley, "Prelude to Antarctic Adventure," SPRI Archives, MS 1097/20/1-3, 2–3.

26. Ferrar, "Report," 9.

27. T. W. Edgeworth David and Raymond E. Priestley, "Geological Observations in Antarctica by the British Antarctic Expedition, 1907–1909," in E. H. Shackleton, *The Heart of the Antarctic: Being the Story of the British Antarctic Expedition, 1907–1909,* vol. 2 (Philadelphia: Lippincott, 1909), 308.

28. Raymond E. Priestley and T. W. Edgeworth David, "Geological Notes of the British Antarctic Expedition," in Congrès Géologique International (1910), *Compte rendu de la XI:e,* vol. 1 (Stockholm: Norstedt & Söner, 1912), 782; E. H. Shackleton, "Some Results of the British Antarctic Expedition, 1907–9," *Geographical Journal* 34 (1909): 492.

29. E. H. Shackleton, "Instructions for the Northern Sledge-Party under the Command of Professor E. David," in Shackleton, *Heart of the Antarctic,* 74.

30. Shackleton, *Heart of the Antarctic,* 28.

31. James Murray, "Report to Shackleton," in Shackleton, *Heart of the Antarctic,* 29; Raymond E. Priestley, "Diary," in ibid., 31–32.

32. Raymond E. Priestley, "Scientific Results of the Western Journey," in Shackleton, *Heart of the Antarctic,* 333.

33. Priestley, "Diary," 62.

34. Bertram Armytage, "Report to Shackleton," in Shackleton, *Heart of the Antarctic,* 68; Raymond E. Priestley, "Diary," Jan. 24, 1909 (3 a.m. entry), SPRI Archives, MS 298/1/8; Armytage, "Report," 69.

35. R. F. Scott, "Plans of the British Antarctic Expedition, 1910," *Geographical Journal* 36 (1910): 12, 17.

36. Frank Debenham, *The Quiet Land: The Diaries of Frank Debenham* (Alburgh, UK: Bluntisham Books, 1992), 41.

37. Ibid., 12; Wilfred Bruce to Kathleen Scott, Feb. 27, 1911, SPRI Archives, 1488/2.

38. T. Griffith Taylor, "The Western Journeys," in R. F. Scott, *Scott's Last Expedition,* vol. 2 (New York: Dodd, 1913), 124.

39. Debenham, *Quiet Land,* 43.

40. R. Scott to Taylor, Jan. 26, 1911, fasc. repr. in Griffith Taylor, *With Scott: The Silver Lining* (New York: Dodd, 1916), 122; Debenham, *Quiet Land,* 45.

41. Taylor, *With Scott,* 149; Debenham, *Quiet Land,* 62.

42. Taylor, *With Scott,* 132; Taylor, "Western Journeys," 130; Debenham, *Quiet Land,* 64.

43. Taylor, *With Scott,* 138; Charles S. Wright, *Silas: The Antarctic Diaries and Memoir of Charles S. Wright* (Columbus: Ohio State University Press, 1993), 88–89; Debenham, *Quiet Land,* 70; Taylor, "Western Journeys," 131.

44. Taylor, *With Scott,* 145.

45. Debenham, *Quiet Land,* 73; Wright, *Silas,* 106.

46. Debenham, *Quiet Land,* 89.

47. Ibid., 125.

48. Taylor, *With Scott,* 344; Taylor, "Western Journeys," 161; Taylor, *With Scott,* 340.

49. Taylor, *With Scott,* 358.

50. Ibid., 386.

51. Tryggve Gran, *The Norwegian with Scott: Tryggve Gran's Antarctic Diary, 1910–1913* (London: Her Majesty's Stationery Office, 1984), 167; Taylor, "Western Journeys," 186; Taylor, *With Scott,* 405.

52. Debenham, *Quiet Land,* 139.

53. E. L. Atkinson, "The Last Year at Cape Evans," in Scott, *Last Expedition,* 221.

54. F. Debenham, "The Geological History of South Victoria Land," in Scott, *Last Expedition,* 295.

55. Ibid., 296–99; A. C. Seward, "Antarctic Fossil Plants," in British Museum (Natural History), *British Antarctic ("Terra Nova") Expedition, 1910, Natural History Report: Geology,* vol. 1 (London: British Museum, 1914), 42.

56. Ibid., 299–300; Griffith Taylor, "A Résumé of the Physiography and Glacial Geology of Victoria Land, Antarctica," in Scott, *Last Expedition,* 286.

CHAPTER 8. THE MEANING OF ICE

1. James Cook, *A Voyage towards the South Pole and round the World,* vol. 1 (London: W. Strahan, 1777), 42; James Clark Ross, *A Voyage of Discovery and Research in the Southern and Antarctic Regions,* vol. 1 (London: Murray, 1847), 221; Clements R. Markham, "Address to the Royal Geographical Society," *Geographical Journal* 41 (1900): 10.

2. Alfred, Lord Tennyson, "Sir John Franklin," in *The Poetic and Dramatic Works of Alfred, Lord Tennyson,* ed. W. J. Rolfe (Cambridge: University Press, 1898), 487.

3. Edward Wilson, *Diary of the "Terra Nova" Expedition to the Antarctic, 1910–12* (New York: Humanities Press, 1972), 212; Alfred, Lord Tennyson, *In Memoriam*, in Rolfe, *Poetic Works of Tennyson*, 176.

4. Alfred, Lord Tennyson, *Ulysses*, in Rolfe, *Poetic Works of Tennyson*, 88–89; Fridtjof Nansen, "To the North Magnetic Pole and through the North-West Passage: Discussion," *Geographical Society* 29 (1907): 514; Markham, "Address," 13; Tennyson, *In Memoriam*, 176.

5. "Glacier," *Encyclopaedia Britannica*, 9th ed., vol. 10 (New York: Werner, 1898), 629.

6. J. W. Gregory and T. G. Bonney, "Ice Observations," in *The Antarctic Manual for the Use of the Expedition of 1901*, ed. George Murray (London: Royal Geographical Society, 1901), 192, 194.

7. Ibid., 198–99.

8. J. W. Gregory, "The Work of the National Antarctic Expedition," *Nature* 63 (1901): 612; Gregory and Bonney, "Ice Observations," 200.

9. Gregory and Bonney, "Ice Observations," 196; Louis Bernacchi, "Topography of South Victoria Land," *Geographical Journal* 17 (1901): 492.

10. Joseph Hooker, "Topography of South Victoria Land—Discussion," *Geographical Journal* 17 (1901): 492; E. H. Shackleton, "To the Great Barrier," *South Polar Times* 1 (August 1902): 41.

11. Gregory and Bonney, "Ice Observations," 193–94.

12. Duke of Argyll, in "The Royal Society's Antarctic Meeting," *Geographical Journal* 11 (1898): 419–20.

13. Edward B. Poulton, "The National Antarctic Expedition," *Nature* 64 (1901): 83.

14. Edward Wilson, *Diary of the "Discovery" Expedition to the Antarctic Regions, 1901–1904* (New York: Humanities Press, 1967), 89; Robert F. Scott, *The Voyage of the "Discovery,"* 2 vols. (New York: Scribner's, 1905), 1:118–19; Shackleton, "To the Barrier," 41; Scott, *Voyage of the "Discovery,"* 1:119.

15. Albert B. Armitage, *Two Years in the Antarctic: Being a Narrative of the British National Antarctic Expedition* (London: Arnold, 1905), 35; Robert F. Scott, "Sea Ice," *South Polar Times* 1 (August 1902): 15.

16. Scott, *Voyage of the "Discovery,"* 1:121; H. T. Ferrar, "Report on the Field-Geology of the Region Explored during the 'Discovery' Antarctic Expedition, 1901–4," in National Antarctic Expedition (NAE), *Natural History: Geology*, vol. 1 (London: British Museum, 1907), 55.

17. Scott, *Voyage of the "Discovery,"* 1:177, 2:417.

18. Ibid., 2:416, 420–21; Armitage, *Two Years in the Antarctic*, 182.

19. Scott, *Voyage of the "Discovery,"* 2:301.

20. Ibid.; L. C. Bernacchi, *Saga of the "Discovery"* (London: Blackie, 1939), 66, 101, 103–4.

21. Scott, *Voyage of the "Discovery,"* 2:299–300.

22. Ibid., 2:259.

23. Ibid., 2:254, 261.

24. William Lashly, *Under Scott's Command: Lashly's Antarctic Diaries* (London: Gollancz, 1969), 75; Robert F. Scott, "Antarctic Glacier Exploration: An Address Delivered before the Alpine Club, March 7, 1905," Scott Polar Research Institute (SPRI) Archives, MS 366/16/17, 7.

25. Scott, *Voyage of the "Discovery,"* 2:261–62.

26. Ibid., 2:260–61, 265.

27. Ibid., 2:277, 279, 282, 284.

28. Robert F. Scott, "Précis of Lecture at Albert Hall, March 7, 1905," SPRI Archives, MS 366/16/15, 3; H. T. Ferrar, "Report on the Field-Geology of the Region Explored during the 'Discovery' Antarctic Expedition, 1901–4," in NAE, *Natural History: Geology*, 66, 73, 99.

29. Scott, *Voyage of the "Discovery,"* 2:423.

30. Ibid., 2:407.

31. T. W. Edgeworth David and Raymond E. Priestley, *British Antarctic Expedition 1907–9 Reports on the Scientific Investigations: Geology,* vol. 1 (London: Heinemann, 1914), ix; Raymond E. Priestley and T. W. Edgeworth David, "Geological Notes of the British Antarctic Expedition," in Congrès Géologique International (1910), *Compte rendu de la XI:e,* vol. 1 (Stockholm: Norstedt & Söner, 1912), 790; E. H. Shackleton, "Some Results of the British Antarctic Expedition, 1907–9," *Geographical Journal* 34 (1909): 483, 494.

32. Priestley and David, "Geological Notes," 791.

33. E. H. Shackleton, *The Heart of the Antarctic: Being the Story of the British Antarctic Expedition, 1907–1909,* vol. 2 (Philadelphia: Lippincott, 1909), 55; James Murray, "The Scientific Work of the British Antarctic Expedition of 1907–9," *Geographical Journal* 36 (1910): 203; Shackleton, "Some Results," 499.

34. T. W. Edgeworth David and Raymond E. Priestley, "Geological Observations in Antarctica by the British Antarctic Expedition, 1907–1909," in Shackleton, *Heart of the Antarctic,* 301.

35. Edward R. G. R. Evans, *South with Scott* (London: Collins, 1921), 1.

36. Victor L. A. Campbell, "Narrative of the Northern Party," in R. F. Scott, *Scott's Last Expedition,* vol. 2 (New York: Dodd, Mead, 1913), 86; Raymond E. Priestley, *Antarctic Adventure: Scott's Northern Party* (Melbourne: Melbourne University Press, 1974), 255, 229.

37. Priestley, *Antarctic Adventure,* 226; Campbell, "Narrative," 90; Priestley, *Antarctic Adventure,* 246–47.

38. Victor Campbell, *The Wicked Mate: The Antarctic Diary of Victor Campbell,* ed. H. G. R. King (Alburgh, UK: Bluntisham Books, 1988), 136.

39. Priestley, *Antarctic Adventure,* 288.

40. Campbell, *Wicked Mate,* 132.

41. Evans, *South,* 282. Similar comments are in E. R. G. R. Evans and H. L. L. Pennell, "Voyages of the *Terra Nova,"* in Scott, *Last Expedition,* 275.

42. Campbell, *Wicked Mate,* 155; Campbell, "Narrative," 91; Priestley, *Antarctic Adventure,* 321, 277.

43. Priestley, *Antarctic Adventure,* 351.

44. T. Griffith Taylor, "A Chapter of Antarctic History," *South Polar Times* 3 (June 1911): 13, 15.

45. Griffith Taylor, "A Résumé of the Physiographic and Glacial Geology of Victoria Land, Antarctica," in Scott, *Last Expedition,* 288; Griffith Taylor, *With Scott: The Silver Lining* (New York: Dodd, 1916), 395 (quoting diary); Taylor, "Résumé," 294.

46. Taylor, *With Scott,* 9, 14; R. E. Priestley and C. S. Wright, "Some Ice Problems of Antarctica," in *Problems of Polar Research,* ed. W. L. G. Joerg (New York: American Geographical Society, 1928), 335.

47. Charles S. Wright, *Silas: The Antarctic Diaries and Memoir of Charles S. Wright,* ed. Colin Bull and Pat Wright (Columbus: Ohio State University Press, 1993), 83.

48. Robert Falcon Scott, *Journals: Captain Scott's Last Expedition,* ed. Max Jones (Oxford: Oxford University Press, 2005), 303; Wright, *Silas,* 141.

49. Wright, *Silas,* 134.

50. Charles S. Wright, "Notes on Ice Physics," in Scott, *Last Expedition,* 306–7.

51. C. S. Wright and R. E. Priestley, *British ("Terra Nova") Antarctic Exploration, 1910–1913: Glaciology* (London: Harrison, 1922), 402–3.

52. Scott, *Journals,* 288.

53. Roland Huntford, *Scott and Amundsen* (London: Hodder), 402; Frank Debenham, *The Quiet Land: The Diaries of Frank Debenham* (Bluntisham, UK: Bluntisham Books, 1992), 120; Scott, *Journals*, 285–86, 291.

54. Wright, *Silas*, 221–22.

55. Ibid., 214.

56. Scott, *Journals*, 365–66.

57. Tryggve Gran, *The Norwegian with Scott: Tryggve Gran's Antarctic Diary, 1910–1913* (Greenwich: Her Majesty's Stationery Office, 1984), 200.

58. Edward Wilson, *Diary of "Terra Nova,"* 229.

59. Scott, *Journals*, 376; Henry R. Bowers, "Diary," "From the South Pole" [Jan. 17, 1912], SPRI Archives, MS 782/6; Wilson, *Diary of "Terra Nova,"* 232.

60. Scott, *Journals*, 376 (corrected to Scott's original text), 378.

61. Wilson, *Diary of "Terra Nova,"* 236.

62. Scott, *Journals*, 399, 405–6; G. C. Simpson, *Scott's Polar Journey and the Weather* (Oxford: Clarendon Press, 1926), 30; Susan Solomon, *The Coldest March: Scott's Fatal Antarctic Expedition* (New Haven: Yale University Press, 2009), 292–98; Scott, *Journals*, 421.

63. Wilson, *Diary of "Terra Nova,"* 238; Scott, *Journals*, 390, 406.

64. Scott, *Journals*, 387 (corrected to Scott's original text); Wilson, *Diary of "Terra Nova,"* 240.

65. Scott, *Journals*, 392; Wilson, *Diary of "Terra Nova,"* 241.

66. L. E. G. Oates, Diary, Feb. 12, 1912, in Huntford, *Scott*, 521; Wilson, *Diary of "Terra Nova,"* 243.

67. Scott, *Journals*, 408, 410.

68. Ibid., 419, 422.

69. Wright, *Silas*, 345.

70. E. L. Atkinson, "The Last Year at Cape Evans," in Scott, *Last Expedition*, 237.

71. Edward Wilson, Field Notes, in A. C. Seward, "Antarctic Fossil Plants," in British Museum (Natural History), *British Antarctic ("Terra Nova") Expedition, 1910, Natural History Reports: Geology*, vol. 1 (London: British Museum, 1914), 6; F. Debenham, "The Geological History of South Victoria Land," in Scott, *Last Expedition*, 300; Seward, "Antarctic Fossil Plants," 42.

72. Debenham, *Quiet Land*, 175.

73. Ibid., 10.

EPILOGUE

1. Clements R. Markham, "A Tribute to the Heroes," *Times* (London), Feb. 17, 1913, 6; Clements R. Markham, "Robert Falcon Scott," *Geographical Journal* 41 (1913): 220; King George V, in "Messages of Condolences," ibid., 220; Queen Alexandra, in ibid., 220; Earl Curzon of Kedleston, "Address," in ibid., 211.

2. "Scott's Message," *Times* (London), Feb. 12, 1913, 7.

3. Roald Amundsen, *The South Pole: An Account of the Norwegian Antarctic Expedition in the "Fram," 1910–1912*, vol. 1 (London: Murray, 1913), 44; Hugh Robert Mill, "Amundsen's 'South Pole': A Review," *Geographical Journal* 41 (1913): 149.

4. "Appreciation by Other Explorers," *Times* (London), Feb. 11, 1913, 10; Sven Hedin, in "Messages and Tributes," ibid., Feb. 13, 1913, 6; "The British Antarctic Expedition," *Nature* 90 (1913): 650.

5. "Message from Captain Scott," *Times* (London), Feb. 11, 1913, 8; "The Antarctic Tragedy," *Spectator*, Feb. 15, 1913, 263; Markham, "Scott," 219; "Self-Sacrifice and Tragedy," *Contemporary Review* 53 (1913): 433.

6. Stephanie Barczewski, *Antarctic Destinies: Scott, Shackleton and the Changing Face of Heroism* (London: Hambledon, 2007), 123. In addition to Barczewski's wide-ranging book, an in-depth analysis of the immediate public response to news of the Polar Party's fate appears in Max Jones, *The Last Great Quest: Captain Scott's Antarctic Sacrifice* (Oxford: Oxford University Press, 2003).

7. "Lord Curzon's Tribute to the Explorers," *Times* (London), Feb. 25, 1913, 6.

8. "Memorial to Captain Oates," *Times* (London), Oct. 27, 1913, 11.

9. "Killed in Action," *Times* (London), Feb. 12, 1913, 6.

10. "Commander Evans at Queen's Hall," *Times* (London), June 5, 1913, 5.

11. E. L. Atkinson, "E. L. Atkinson's Report of the British Antarctic Expedition 1910–13," 1 (pt. 1, sec. a), Scott Polar Research Institute Archives, MS 280/28/1, 33–34.

12. "The Death of Sir Clements Markham," *Times* (London), Feb. 2, 1916, 5.

13. Barczewski, *Antarctic Destinies*, 260.

14. E. J. Hobsbawm, *The Age of Empire, 1875–1914* (New York: Vintage, 1989), 13.

INDEX